EXPLORING NEW FRONTIERS IN POLICING: SYSTEMIC CONDITIONS, STRESS AND COPING, NEUROSCIENCE, AND ALTERNATIVE POLICE TRAINING

I0478047

WRITTEN BY DR. JOHN "JAY" HALL

TABLE OF CONTENT

ABSTRACT

A copy is a reproduction of an original. Repeating the same process produces the same results. Human beings are different from inanimate objects. They are not carbon copies. They are originals. They are unpredictable because they have the ability to adapt, change, and recreate themselves. At present, our DNA sequence consist of 3 billion nucleic acids and 25,000 genes. For years, our DNA has adapted to our environment. At some point in our remote future, interplanetary travel may change our body chemistry to adapt to a new planet or environment. If we intent to reach that portal as a time traveler, we need to become more aware of the impact that our current environment has on us.

This article looks at the systemic conditions that contribute to some of the negative decision choices that we as members of society make. This article seeks a constructive dialogue on why these systemic conditions continue to exist over decades. One explanation is that society tends to focus on the flaws of the individual rather than the discrepancies associated with some environment(s) that may contribute to deviant behavior. In this article, the author uses police shootings as a catalyst to examine a broader set of systemic conditions that impact criminal justice reform efforts and practices. The author cites the school to prison pipeline as an illustration of a set of dysfunctional systemic conditions that continue to plague minority communities. These environmental conditions also impact the quality of life of police officers as well.

Due to the longevity of certain environmental conditions, the author ask an ethical question, *in lieu of the evidence, what are our goals*. By asking this question, the author believes that criminal justice reform advocates needs to re-examine their intentions. This author proposes a simultaneous top down and bottom up intervention approach to address systemic conditions and to suggest alternative police trainings. The top down approach addresses the ecological, social, economic, educational, and political factors which the author refers to as "supply side policing" factors which shape the environment in many communities of color. In addition, the author suggest a bottom up approach in the form of neuroscience which can be used to "self-regulate" the *somatic markers* that police and citizens develop as a result of interacting with their environment. The TD-BU interventions aim to address the negative quality of life issues that both police and citizens experience. But more important, these interventions will ensure future generations their birthright to thrive rather than just survive in a changing world that demand every aspect of their growth and development.

PART I. SYSTEMIC CONDITIONS:

Welfare, Punishment, and Public Policy (1)

Our environment is a representation of what we have created. It is a tapestry of public and private policies reflecting political, economic, educational, and social philosophies. Periodically, we take an assessment to determine whether past, present, or future policies will support or detract from our vision of what we expect society to look like. Despite our understanding of the role that policies play in creating our environment, there has been a refusal to recalibrate in order to adapt to changes when unintended or unanticipated consequences occur. This failure to change in the face of compelling evidence has resulted in a number of environmental conditions that not only affect present generations but our future generations. One of the qualities that differentiate human beings from computers is pattern recognition. Our brain uses its 100 billion neurons, at the same time, to retriever similar events and or experiences for us to evaluate. Some of the examples below represent environmental conditions that have been ignored or deferred resulting in a set of systemic conditions that reign inequalities among the least of us.

For example, in a study on welfare and punishment, Downes and Hansen (2005) wrote: "states with less generous welfare spending having higher imprisonment rates in the 1990s, but this later period also saw states with a higher proportion of blacks, other ethnic minorities and greater poverty having higher imprisonment rates" (p.8). Despite these correlations or causal relations between crime and poverty, corrective social policy initiatives and or equitable financial spending in areas where obvious **inequities** do exist have not changed much. Another example closely related to welfare funding and prison funding includes funding for public education. In most states, our education spending trails correctional spending.

This author is concern with why these conditions persist. In an attempt to answer this question, this author points to two concerns: a) **how we frame or define a problem** and b) **our intentions with respect to aligning the goal, the plan, and the results.** There is also a concern with respect to how each discipline views the same problem and how each discipline is interdependent upon the other. For example, there is a distinct difference between how public organizations view a problem in comparison to how private organizations view a problem. Borrowing two concepts from the research of Burt, R.S. (1996), this author believes that public organizations frame problems from the perspective of human capital; whereas, private organizations frame problems from the perspective of social capital. The examples cited above represent how public organizations place an emphasis on human capital.

According to Burk (1996), human capital refers to an individual's ability; whereas, social capital refers to a social network structure between people and across multiple disciplines. Burk further conceived that inequality can exist in both; however, the inequality exist for different reasons. Burk stated that the inequalities with respect to human capital were a result of the **differences in an individual's abilities.** By contrast, the inequalities associated with social capital were a result of **contextual differences** among people. Contextual differences includes differences in *"resources and opportunities"*. Unfortunately, public organizations usually adopt policies based on a human capital point of view. This approach minimizes the role of the contextual differences that people inherit based on their circumstances. Failing to see a **connection** between the individual's abilities and the individual's need for available resources and opportunities to develop those abilities can be problematic. Its' a step backward to the era of separate and *not* equate.

By comparison, private organizations usually adopt policies from a social capital perspective. Burk informs us that a social capital point of view seeks to build social network structure(s) which add value and opportunities by embracing outcomes that reflect: transformation, change, innovation, diversity, empowering people, sharing of information, sharing a vision, and building trust. The social capital concept is based on developing a social network structure to maintain or acquire a **competitive advantage.** This competitive advantage is what differentiates private organizations (business entities) from public organizations. Because public agencies have no competitive rival, there is no economic incentive to correct bad policies or to change programs that are not working. Because there is no competitive rivals in the public sector, it becomes difficult to draw comparisons to evaluate policy and program outcomes in a timely manner. Therefore, failure, more often than not, is attributed to an individual's lack of abilities, skills, or knowledge. Under the human capital perspective, seldom, do we hold organizations accountable for errand or failed policies. Each problem can be viewed from a number of perspectives; therefore, how we define a problem becomes critical to how we resolve it.

With respect to the second point, the author believes that our intentions, in most cases, reflect the outcome(s) we desire. Our intentions are a test of our integrity. Are we doing what we actually said we were going to do. Seigel (2018) stated, our intentions should be considered first. Seigel (2018) offered that following prescription for understanding and changing our human behavior. The model that he constructed began with what are our intentions. His motto reads, "how our intentions (glows) determines where our attention goes, and how our neural firing flows (positive or negative), and to that degree will determine how our neural and interpersonal connections will grow (or deteriorate)" (p.97). Based on this assessment, our intentions are the most important decision factor in determining the outcome of a problem. Intentions test our integrity and serve as our moral compass to guide all subsequent decisions.

In too many situations, our ethics become compromised by a personal or institutional agenda. In some criminal justice reform situations, politicians often linked social welfare and crime concerns to the public sentiments of the day. And as a result, their administration may adopted crime policies based on whether they wanted to be perceived as being "tough on crime" or "soft on crime". And as a result of these two policy positions, their administration will use their authority to decide the amount of resources and who will receive what resources. Under the human capital perspective, these policy positions convey the message to the public that individuals who are lacking in their abilities are the reason(s) that they are on welfare, commit crimes, or drop out of school. Very seldom do we examine the antecedent conditions that lead up to the behavior(s) in question.

Despite our knowledge (awareness) and our ability to correct errors when policies and programs are not working; sometimes, we still failed to correct them. This author agrees with Seigel (2018) that our intentions not only represent our moral compass but also our goals or our end game. I don't believe any person in leadership will agree that they purposely engineered systemic conditions as their end game. However, they are accountable for not trying to correct the problem. Affixing blame does little to help the people most in need. The purpose of this manuscript is to address the issue (now). Therefore, we must abandon the blame game and the practice of silo thinking. We must begin to integrate, coordinate, and synchronize the concepts of human capital and social capital so that we can begin to eliminate the systemic conditions described above. As a law enforcement practitioner, this author believes that the true litmus test for assessing criminal justice reform efforts begins when we develop futuristic strategies to address these debilitating conditions that continue to adversely impact the quality of life for both police officers and communities of color. In the next section, we look at how the environment (systemic conditions) influence the coping behaviors and relationships of both police and citizens.

The Environment and Our DNA (2)

Today, most researchers would agree that an individual's personality is shaped by both genes and the individual's environment. All organisms, human and otherwise, have two functions which are to survival and to thrive. They are equip with an executive function. For most of us, we have been taught that our brain, alone, was where our executive decisions were made. The role of the executive function is to make sure that the organism is maintaining homeostasis. When an organism gets out of balance; it must try to re-organize itself or it will become dysfunctional. Adapting to our environment is a learning process and is necessary for our self-preservation. People also adapt to stress for similar reasons, some good and some bad. Research has informed us that the role of the environment can have a strong impact on our DNA. In an epigenetic inheritance study of holocaust survivors, Thomson (2015) wrote that the environmental stress associated with living the life of a holocaust survivor changed the brain's chemistry and altered the survivors' DNA which was passed on to their children. Dr. Lara Boyd, a neuroscientist, stated that this is possible because the brain is capable of learning and changing behavior in three ways: a) the release of certain chemicals to impact short term memory, b) structural changes to impact long term memory, and c) functional changes based on new connections in the brain's neural networks (2006). Our environment represents the contextual framework within which our behavior is learned.

When we see a picture of a lion, as opposed, to seeing an actual lion in the wilderness, our bodies respond differently. The difference in behavior can be attributed to context. Context is equally important in police shootings. What an officer perceives at the time of an officer involved shooting, can be the difference between a life or death decision. When we fail to incorporate context within our problem solving equation, we lose a large amount of information. An officer's perception of the environment represents a particular mindset. Siegel (2018) defined mindset as a collection of the following: intentions, attention, awareness, emotions, memory, and patterns of behavior (p.97). Our mind set is influenced by our different life experiences. In support for this point of view, Southwick and Charney (2012) stated that our life experiences are unique to each individual and are influence our ability to adapt and cope with adversity. Southwick and Charney further stated that adapting and coping successfully (with stress, fear, emotions, feelings, anxiety) depends on a complex set of outcomes which include genetic, biological, psychological, social and spiritual factors available for each individual. These individual differences are worth noting in training exercises since most police training is based on the assumption that one type of training fits all. In the following sections, the author describes how certain environments have adversely impacted the quality of life of both citizens and the police.

Ethnic Groups and their Environment (A)

In this section, we borrowed from the Pew Research Center (2013) quality of life indicators reflecting the differences of four basic ethic groups: blacks, whites, Hispanic, and Asians. The research spans over several decades. The purpose of the research is to illustrate the adverse impact that the environment can have on some ethnic groups. The indicators used to compare groups were income, poverty, wealth, education, life expectancy, marriage rates and single moms, and incarceration. With respect to (**income**), whites earned $67, 175 compared to $39, 769 for Blacks, $68, 521 for Asians, and $40, 007 for Hispanics (Census Reports, 2011). In the 2012 Population Survey, Black Americans were found to be nearly three times likely to be in (**poverty**) compared to Whites. Among whites, ten percent (10%) were poor in 2011, as compared to twenty eight percent (28%) of blacks, twenty five percent (25%) of Hispanics, and 12% of Asians. With respect to (**wealth**), white households had a net worth of $91, 405, whereas blacks had a net worth of $6,446, Hispanics had a net worth of $7,843, and Asians $91, 203. Within the same 2012 survey, whites owned their homes seventy three percent (73%) of the time, blacks owned their homes forty four percent (44%) of time, Hispanics owned their homes forty six percent (46%) of the time, and Asians owned their homes fifty seven percent (57%) of the time, as cited in Pew Research Center (2013).

On the topic of (**education**), ninety two percent (92%) of white adults completed high school compared to eighty six percent (86%) blacks, eighty nine percent (89%) Asians, and sixty five (65%) Hispanics (Pew Research Center, 2012). When considering college completion, the pattern changed somewhat. Asians completed college at a rate of fifty one percent (51%) compared to whites at thirty four percent (34%), blacks at twenty one percent (21%), and Hispanics at fifth teen percent (15%). Another significant social indicator by the National Center for Health Statistics (2012) was (**life expectancy**). Whites had a life expectancy of 78.9 years compared to 75.1 years for blacks. Two other important social indicators were (**marriage rates and single moms**). In 2011, Fifty five percent (55%) of white adults 18 and older were married compared to thirty one percent (31%) of black adults 18 and older. In 2011, seventy two percent (72%) of black women who gave birth were unmarried compared to twenty nine percent (29%) of white women who gave birth and were unmarried (National Center for Health Statistic, 2011). The last indicator addresses (**incarceration**). In 2011, white men were under local, state, and federal jurisdiction at a rate of 678 inmates per 100,000 for white U.S. residents. By contrast, black men were under local, state, and federal jurisdiction at a rate of 4, 347 inmates per 100,000 for black U.S. residents. In 2010, the incarceration rate for Hispanics were 1,775 per 100,000 for Hispanic U.S. residents, as cited by Pew Research Center (2013).

These quality of life indicators and environmental factors represent a contextual framework that trigger feelings of hopelessness, depression, and danger for many communities of color. Ironically, some police view their work environment with the same hopelessness and cynicism. One of the biases and stereotypes associated with the citizens in these communities is that they have no aspirations to do any better. Unfortunately, this is not true. Most of the residents in these communities reflect a strong desire to improve their situation. What often is lacking is an accounting of urban policy initiatives which over the years contribute to many of the social disparities listed above. These policy initiatives very seldom are legislated by line officers or neighborhood residents. In the next section, the author examines the same environment or occupational environment, in order to identify some of the stress related coping behaviors that police exhibit.

Police and their Environment (B)

In contrast to the quality of life issues shared by the Pew Research Center data, police organizations are more internal and less transparent. Quality of life issues that police experience are contained within the organization rather than publicized. More often than not, the purpose for this containment is to *protect the image of policing as an institution*. In this section, the author will identify several behavioral outcomes influenced by an environment which is characterized by stress, uncertainty, and danger. Several of these behavioral outcomes include: police shootings, suicide, post-traumatic stress disorders, domestic violence, and heart attacks.

As mentioned earlier, there has been a public outcry for police reform regarding the disproportionate number of minority police shootings; so, we will began, there. From 2015 to 2018, **police shootings** have remained constant. According to Tate, Jenkins, and Rich (2018), there were 995 fatal shooting of citizens by police in 2015, 962 in 2016, 987 in 2017, and 625 in 2018. By comparison, the number of **police officers killed in the line of duty** were: 161 police officers in 2010, 69 police officers in 2011, 49 police officers in 2012, 126 police officers in 2013, 155 police officers in 2014, 160 police officers in 2015, 159 police officers in 2016, 133 police officers in 2017, and so far 73 police officers in 2018 (Police The Law Enforcement Magazine, 2018). During these tragic incidents, police voiced their concern for blue lives as did the minority communities voiced their concerns for black and brown lives. In both situations, the author believes that both the public and the police should begin a dialogue on their "common enemy". Their common enemy is the systemic conditions which influences behaviors that may lead up to an officer involved shooting or any number of maladaptive behaviors.

In addition to police shootings incidents, police and their families are also affected by suicides. According to Walker (2018), there were 140 police officers who committed (**suicide**) in 2017 compared to 108 in 2016. In 2018, there were five Chicago police officers who took their own life. One of the retired police officers whose job was to counsel troubled officers stated that officers are afraid to ask for help because they are stigmatized by their peers, so they continue to carry the stress of the job with them until something unfortunate happens. Another area where little attention is shed on the police culture is post-traumatic stress disorders. With respect to **post-traumatic stress disorder**, Thomas, T.S. (2018) cited a study by the National Alliance on Mental Illness which reported that 7 to 19 percent of the police profession suffer from post-traumatic compared to 3.5 percent of the general population. With respect to the issue of **domestic violence**, Friedersdorf (2014) stated that officers have a two to four time greater rate of domestic violence than the general population. In fact, more police committed domestic violence than professional football players. In a discussion on **heart attacks**, Weaver, D. (2015) reported that ten percent of all police officers deaths can be attributed to heart attacks. Some other quality of life topics that weren't discussed but should be considered in sequent police related research include high divorce rates, alcoholism, drugs, and or excessive gambling.

In the research provided by the Pew Research Center and by the above police sources, both the police and citizens are adversely impacted by the same environment. As a result of various stressors that certain conditions in the environment imposes, police and citizens find different ways of coping and adapting. In most cases, the coping and adapting behaviors by the police and citizens are their response to *an unhealthy environment*. Over time, these coping and adapting behaviors become "hardwired" and "reinforced" because the environment or the systemic conditions have not changed. Why the systemic conditions have not changed is a primary concern of this author. And until, there is a meaningful dialogue on addressing the root causes of the systemic conditions; then, we will continue to see disparities in the quality of life issues among communities of color as well as the police.

In a more recent study including police, fireman, paramedics, correctional officers, nurses, constables, and dispatchers, the qualitative research of Ricciardelli et al (2018) confirmed a number of physical, emotional, and mental ailments are related to with occupational trauma experienced by public safety personnel. According to Ricciardelli et al (2018), these experiences of occupational trauma are "intensified by a **hierarchical**, often paramilitary, **reporting structure** within such organizations that *discourage* PSP (public safety personnel) from expressing any vulnerabilities" (p.568). Coupled with the previous fact that most police feel that they will be stigmatized by their peers for seeking help; then, the role of police leadership in this area becomes critical in terms of leading change for the better. In the next section, the author provides an information processing chart to depict the myriad of factors influencing an officer or individual's perception and possibly their decision making process.

PART II.

THREAT PROCESSING, STRESS IMPACT, AND DECISION MAKING

First, emotions tell us what is good for us and what is harmful. The vast majority of things in the world are either harmful or not very useful. When we experience the emotion of "like," we are learning to identify the tiny fraction of things in the environment that are beneficial to us (Kaku, 2011). When we experience a threatening situation, our emotions alarm us that danger is near. According to Kaku (2011), our brain is a neural network which processes our experiences from the bottom up. As we learn behaviors or make decisions, our brain strengthens the neuron circuitry. New task result in "rewiring"; whereas, repeating old task result in "hardwiring". This process of neural pathway reinforcement is based on Hebb's law, where neural pathways are reinforced by the extent of their use. The *reinforcement of neural pathways* explains why some behaviors are difficult to change (Kaku, 2011).

Because we don't learn emotions in a vacuum, there is a social aspect of our decision making. The individual or group has norms or rules that they have to follow which includes the group's ethics and moral values. In law enforcement, we address this same concern in the form of a suspect's culpability. Kolhberg (1981) identified six moral development stages that included self-interest egos, group and social norms, legal, political, and universal principles. Because human beings are social creatures, many behaviors are sanctioned by the *group's norms and decides are made based on the **groups' moral code***. This does not necessarily mean that all the group's norms are universally accepted. Group norms reinforce the notion that human beings like to fit in and that some behavioral decisions are indeed influenced or filtered by peer pressure or group think.

Understanding how we process information can be a key in understanding our decision making. The author list an information processing chart. The chart includes three major components: the knowledge domain, the decision making skill domain, and the executive process domain. Starting with our perception of a stimulus or stimuli, our brain begins a process of encoding these events, experiences, and or situations. Encoding involves changing chemicals into electric impulses in the brain. These impulses are first scanned by our short term memory. Our short term memory determines within 1 to 3 seconds if something is important or should be forgotten. If important, the information is retained and transferred to our working memory for a duration of about 5-15 seconds before forgotten. Finally, experiences are stored in the long term memory based on *the meaning that an individual gives to a particular event or situation* (Dataworks, 2014). The chart includes four sensory encoders (visual, acoustic, elaborative, and semantic) to sort information about the stimuli received. In visual encoding, the amygdala assigns positive or negative emotional values to the stimuli which in some cases can reinforce memory retention (Salvage, 2018). At the same time that we are encoding information, we are also **filtering** the stimuli information based on our own past experiences or the vicarious experiences of others. After that, we apply decision making rules and make an executive decision. **See chart below**:

INFORMATION PROCESSING FLOW CHART

Environment   **Event and Experience**   **Encoding of Event and Experience by Brain**

Maclean Brain Model: **Instincts – Emotions - Cognitive**

Sensory Encoding Lenses:

Visual Images   Acoustic   Elaborative (Link New to Old)   Semantic

---

Knowledge Domain (Content/Individual Files)

Filters:

Environment, DNA, Culture, SKAs, Moral Development, Cynicism, Mental Model, Self-Awareness, Emotions, Feelings, Spirituality, Beliefs, Assumptions, Mental State, Mindset, Patterns of Behavior, Procedural Memory, (Somatic Markers), (Hypervigilance), Critical Thinking, Implicit and Explicit Biases, Stereotypes, World View

---

Sensory Memory (Threat Detection Criteria - Pleasure v. Pain)

Short Term Memory (Characteristics – 20 seconds, seven independent items, distractions)

Long Term Memory (Semantic Memory –Fact Base and Episodic Memory –Context Base)

Decision Making Skills Domain (Software/Hardware)

Training, Legal Guide Lines, Cultural Norms, Ethics, Loyalty, Organizational Politics

---

Executive Process Domain (Computer Processing)

Associative Principle of the Brain (Retrieving Similar Events)

Reactive Learning (Habitual Downloading Past Behavior)

Decision Making Process (Communication, Analysis, Synthesis, Prioritizing, Decision)

Performance Outcomes: Experiences Re-Deposited Back Into Community

---
The Perception of How Police Performed: Builds Trust or Undermines Trust

---

Information Process Chart modified by Hall (11/21/18)

Our Reticular Activating System (3)

In addition to the information processing that normally occurs, police officers are socialized by an *occupational culture* which teaches them not to express their emotions or their feelings. For similar reasons, police are taught to behave in a macho persona because it supports a strong police image. In conjunction with this macho image, police have adapted a "warrior" mentality. Most criminal justice practitioners will agree that the "warrior" mentality encourages a "we" versus "them" mindset that drives a trust wedge between the police and the communities that they serve. Nonetheless, the warrior mentality is part of the police cultural values. In describing police culture, Gilmartin (1986), a police psychologist, stated that police as a result of their environment have developed a state of hypervigilance that results from their *concern for officer safety*. Gilmartin (1986) further indicated that this constant state of awareness alters the neural circuitry of an individual and can lead to being over-reactive to threatening situations. At a bio-behavioral level, Gilmartin stated that the affected neural circuitry is the **reticular activating system**. This system scans inputs from an officer's perceptual field and then determines which events should be interpreted as threatening (Gilmartin,1986).

In exercising their daily routines, the average citizen may be oblivious to threats in their environment. However, by comparison, police are trained to scan for any innocuous event for the possibility of danger. For police officers, the *threat threshold* becomes exponential. In order to learn new ways of behaving, Gilmartin stated that a **new perceptual set** would require that the reticular activating system be taught a *new set of values* for interpreting incoming (social) cues and putting valances of potential danger on events the average citizen would clearly interpret as neutral (Gilmartin, 1968).

This point by Gilmartin is very significant for two reasons: a) the "warrior" mentality represents a cognitive bias learned and reinforced by the police culture, and b) a new set of values must be created to support the transformation of a new mindset. Under the Obama Administration, the task force articulated a new set of values to transition police culture from a "warrior mentality" to a "guardian mentality". Unfortunately, such a transformative change takes time and requires uniformity in values and practices. A key factor in addressing a new set of values is to assess the role that the current environment (systemic conditions) have on police values such as the "warrior" mentality. Without making a critical assessment of the causal relationship between a high crime neighborhood and the "warrior" mentality, we miss an opportunity to correctly define the problem and develop an effective solution.

In general, Amodio (2014), stated that our prejudices (our biases) are learned behaviors that drive us to discern an "us" versus "them" mentality. Prejudices are based on a preference for one's own kind (familiarity, safety, security) and therefore, *outsiders* are viewed as a threat (Amodio, 2014). Our perception that an individual or group is threatening; also, has some neuroscience and hormonal effect. The hormone, oxytocin, which functions in social bonding, can influence our degree of loyalty toward in-group members or can increase our tendencies toward negativity, anger, aggression, or violence toward out-group members. While the state of hypervigilance cited by Gilmartin reflects a cognitive bias based on the police occupational culture, it centers around how and what police perceive as a threat.

In describing the reticular activating system, Schneider (2017) stated it was a bundle of nerves that are positioned in the brain stem which are used to *scan the environment and filter out those things that a person feels are not important. It can also be used to filter in and filter out information to validate our belief system.* Schneider (2017) stated, **"it filters the world based on the parameters you give it"** (p. 2). This point is very significant because it reinforces the fact that our (limited) experiences can distort how we see the behaviors of others as we view them from our own experiences. When we fail to recognize our biases, we run the risk of distorting the meaning of information or ignoring crucial information that may improve our assessment or problem solving capacity.

Because police develop a cognitive bias that originates from their occupational culture, their perception may be different than the average citizen. However, culture is not the only factor that may influence one's perception. One's experiences with race and gender can also influence an individual's decision making. We see how subjective experiences influence the response of officers on the same department. We see this scenario play out culturally when the cultural awareness of a black officer allows him or her to decode a situation as non-threatening; where, a white officer may view the same situation as threatening.

Without testing how individual differences based on DNA, emotional histories, parenting, sibling positioning, poverty, education, race, culture, stress, class, and status **impact** an officer's performance, we have no way of predicting what an officer thinks or how they will perform under stress. One seminal research study conducted by Heider and Simmel (1944) bears some light on how our perceptions work. According to Heider and Simmel, research participants were told to interpret the movement of geometric shapes such as triangles, squares, rectangles, and lines. What the researchers discovered, without any social cues, was that the participants framed the movement of shapes in the social context of people and began to assign intentions and goals to geometric shapes as if they were real people.

Applying this research to police simulation training raises several questions. Based on Heider and Simmel research, *can police simulation training act as a social cue that stimulates an officer's perception before arriving at the scene of an emergency.* This is a question that lends itself to the mental and or emotional state of the officer prior to running the call for service and whether or not the officer's perception influence his or her behavior. In the next section, the next researcher shares his findings on the physiological effects that the body experiences under stressful combat situations and some acceptable performance thresholds.

Physiological Impact of Stress, Fear, or Danger (4)

When threatened in a combat situations, our body responds differently than when we are not stressed or responding to a dangerous situation. Under stressful situations, our heart beat increases, our bronchial passages increase, we perspire, our blood pressure rises, and our ability to digest food slows down in order to allow our body to prepare to act. In general, a threat can be any situation where we detect a change in a known set of routines or patterns. Change triggers an alarm. This process of detecting changes in routine patterns is normal. In police work, a small change can result in a deadly outcome, such as an officer involved shooting.

In his book, Sharpening The Warrior's Edge, Bruce Siddle (1995) discussed how the body is affected during combat situations. Siddle based his initial research on how soldiers performed under stress. He compared the sympathetic nervous system's effect on vital organs between combatants and non-combatants. In researching the two groups, he measured their survival stress reaction by examining their heart beats per minute, visual depth perception changes, eye dilation, auditory changes, motor skills, and brain recall. Below are some of the survival stress reactions (SSR) that he found impacting the **body:**

Heart

At 115 beats per minute (bpm), most people lose line complex motor skills such as finger.

Dexterity

At 145 beats per minute (bpm), most people lose complex motor skills.

Visual Effects

The visual system is the primary sensory organ of the body. At 175 beats per minute (bpm), pupils dilate and flatten resulting into *"tunnel vision"*. A person backs up to gain addition information on the threat. Studies cite a 70 percent decrease in their visual field.

Auditory Effects

At approximately 145 beats per minute (bpm), that part of *the brain shuts down and hearing becomes difficult*. The recall problem is called "Critical Stress Amnesia". At approximately 185-220 beats per minute (bpm), individuals may become paralyzed, show irrational behavior, or the deer in the headlights syndrome.

In conclusion, Siddle stated that the higher the heart rate, the more stressful encounters will affect one's *perception of the threat*. He also noted that in combat situations a person's heart rate can go from 70 beats per minute to 220 beats per minute in less than half a second. *By keeping the heart rate between 115 to 145 beats per minute, Siddle suggested that an officer should still be capable of performing adequately.* Today, there are a number of breathing exercises and mediation techniques to calm the body. Biometric testing research should be factored into shoot, don't shoot training scenarios. When threatened or under stress, our body releases chemical hormones such as norepinephrine and epinephrine to increase the body's capacity to cope with an alarming situation. Another hormone that the body releases is cortisol. During stressful encounters, cortisol is used to process glucose, control inflammation, and control blood pressure (Healthdirect, 2018). However, cortisol, at elevated levels, can increase the risk of depression, mental illness, and lower life's expectancy (Bergland, 2013).

Cortisol the Silence Killer (5)

According to Charlesworth and Nathan (1982), cortisol can also interfere with productivity, learning, and interpersonal relations (p.7). From a wellness point of view, police officers are required to be hypervigilant on a constant basis. These prolonged levels of stress can wear down an officer's immune system. Dr. Hans Selye, the "father of stress research", who coined the "fight or flight" term, developed a model describing how the immune system is weaken by stress. According to his general adaptation syndrome model, the first stage is the *alarm stage* where an event triggers the fight or flight decision. Secondly, there is the *resistance stage* where the alarm signal lessens and the immune system becomes weak. And third, there is the *exhaustive stage* where the exposure to the stressor will finally destroy the organism's energy to resist and can result into illness or death (Charlesworth and Nathan , 1982, p.9).

Cortisol is not only detrimental to adults, it also affects the health of children. In a 2015 study, entitled "Tracing Differential Pathways of Risk: Associations Among Family Adversity, Cortisol, and Cognitive Functioning in Childhood," researchers found that cortisol levels, among children who lived in poverty, an unstable home environment, or homelessness, adversely impacted the child's social–emotional adjustment, as well as, language, motor, and cognitive functions. These findings also included environments where domestic violence were present. The researchers found that children with relatively high or relatively low levels of cortisol are more likely to experience learning deficiencies and have cognitive development issues**,** *as cited in* Psychology Today by Bergland (2015).

This vulnerability of children has not gone unnoticed. In Obama's second inauguration speech, he declared, beyond talk of gun violence legislation, if our legislators and business leaders strive to create policies and fund initiatives that create social connectivity among at-risk teens and reduce **bullying**, they will be reducing cortisol levels in our young people. This will make them mentally and physically healthier, more resilient, and less likely to be **violent**, as cited in Bergland (2013). Given the facts and knowledge that is available to us, this author believes that we have an ethical obligation to address the systemic conditions that negatively impact the quality of life issues that police, citizens, and children face. These two sections on the physiological impact of stress on the body in combat situations and how environmental stressors produces cortisol are further evidence that our environment can be deleterious to our health. In the next section, one police department headed for positive change is the Houston Police Department which commissioned a study on officer involved shootings.

Officer Involved Shooting Project (June 4, 2014) (6)

In 2014, the Houston Police Department along with Northwestern University conducted a study on officer involved shootings. The study was entitled, Mathematical Methods in The Social Sciences Houston Texas Police Department on Officer-Involved Shooting. The study was a proactive approach to reducing officer involved shootings.

This particular study was designed to statistically answer the following types of research questions: a) what was the officer's call history for that shift prior the officer involved shooting, b) what type of information about the event did the officer have before arriving at the location, c) what type of verbal exchange occurred between the officer and the suspect, d) how did the issue of cover and concealment affect the outcome of the scene, e) what physical actions were taken by the suspect prior to the discharge of the officer's firearm, f) what reactions did the officer take in response to the suspect's actions, g) what type of tactical training did each officer have prior to being involved in a shooting incident. The study's data was collected from HPD case files from 2005 to 2013.

While the overall findings are significant in building a theoretical de-escalation model, the study was designed to answer the *"what"* questions: what procedures influenced officer involved shootings. In this author's opinion, the variables that seem to carry the most weight were verbal commands, conceal and cover practices, and whether or not officers' had guns drawn prior to an officer involved shooting situations. Findings on the verbal command data revealed, there was a probability of 0.918 percent that deadly force would be used if verbal commands were not used compared to 0.591 when verbal commands were made. In the study, the top three verbal commands used along with their percentage breakdown were a 0.297 for stop, 0.231 for drop the weapon, and 0.231 for show your hands.

Since officers were not interviewed in the study, it was very difficult to link the officer's verbal commands to the officer's perceived threat assessment. Data was also collected on cover and concealment procedures. Researchers used offense reports to examine cover and **concealment** practices. The analysis revealed that there was a 0.297 probability for possible concealment for officers and a 0.287 probability for possible cover for officers. In terms of actual officer concealment, the probability was 0.224 percent. For actual cover, the probability was 0.211 percent. Again, without interviewing each officer, it was very difficult to conceptualize an officer's threat assessment as it related to cover and concealment practices at the scene.

Another area of concern was the percentages of incidents where officers had their **gun drawn.** Based on the study's findings, thirty seven percent (37%) of officers with gun drawn were more likely to use them, nineteen percent (19%) of officers who knew that the suspect had a weapon was more likely to use their gun, and twenty three percent (23%) of officers riding solo were more likely to use their weapon compared to a two man unit. Another area of concern, the study focused on was the behavior of the suspect. Based on the study's findings, suspects who displayed an **aggression stance** were thirty five percent (35.1%) more likely to provoke an officer involved shooting as compared to twenty two percent (22.4 %) when a suspect didn't display an aggressive stance. Again, without interviewing the officer(s), questions regarding an officer's perception before or at the time of engagement are difficult to answer. Overall, the study makes a major contribution to de-escalation research by identifying key police procedural practices relative to officer involved shootings.

Based on the study's research, eighty two percent (82%) of officers were less likely to use their guns when responding to critical incident training situations as a result of critical incident training. The verbiage that was used to cite this findings was not clear so I felt it inappropriate to comment. On the other hand, Colucci, McCleary, and Ng, (2014), the research designers, pointed out that while there is extant research on the traumatic effects that OIS shootings have on the emotional state of police officers; there is little or no study that has examined the inverse relationship. **This inverse relationship should ask how does an officer's emotional deposition contribute to an officer involved shootings.** In support of this concern, Thomas J. Aventi stated that "the nature of police shootings elicits considerable emotion, as well as ample (doses) of misinformation" (Aventi, 2003, p.1), as cited in Colucci, McCleary, and Ng. While on-scene conditions directly affect the likelihood of gun use, it is worth studying the impact of **pre-scene conditions**, such as an officers' emotional state, as well, because they also play a role in determining the context and the scene leading up to the eventual gun use (Colucci, McCleary, and Ng, 2014, p.7).

While one of the objectives of the above research was to limit the number of officer involved shooting (Colucci, McCleary, and Ng, 2014, x-xi), this objective is difficult to achieve unless additional research is provided on **an officer's emotional state of mind** as part of the contextual framework relative to a shooting incident. This **(data)** is essential if we intent to reduce the number of officer involved shootings. In fact, this requirement has recently become a FBI mandate regarding OIS. In the next section, the author focuses on the role that emotions-feelings play in influencing behavior via neuroscience.

PART III NEUROSCIENCE:

Executive Decision Debate: The Cognitive Brain or the Emotional Brain (7)

Our emotions, feelings, and state of mind, are several of the many filters listed that influence our decision making process. Paul MacLean (1961) used one model to illustrate the evolutionary development of the human brain. According to MacLean, first, there was the reptilian brain which was characterized by our (instincts) which governed our basic body functions. Then, there was the limbic brain which was possessed by the first mammals. The limbic brain included the hippocampus, the amygdala, and the hypothalamus. The amygdala is better known as the (emotional) brain. Next came, the neocortex or the (rational) brain which developed last but is the largest portion of our brain. The neocortex contributes to learning, language, abstract thinking, complex problem solving, and creativity. On an evolutionary scale, the **reptilian** brain first appeared in fish, nearly 500 million years ago (McGill University, C.A. 2005). Years after the tribune model was conceived by MacLean, other researchers opined that the MacLean model was an over simplification of the brain's structure because many functions of the brain are integrated in each different part (Gould, 2003). In the describing the reptilian brain, Kaku (2011) stated that its' primitive life functions were balance, aggression, territoriality, searching for food, etc. In comparison to the reptilian brain, the **limbic system** first appeared in small mammals, about 150 million years ago. And last, the **neo-cortex** began to develop in primates, some 2 or 3 million years ago, as the genus *Homo* emerged (McGill University, C.A. 2005).

According to Kaku (2011), emotions played a major role in socializing animals to the rules of the pack by rewarding or punishing certain behaviors. These rules of the pack were necessary for the survival of the species. Kaku (2011) further stated that without emotions, everything has the **same value**, and we become paralyzed by endless decisions. Emotions help us prioritize what's important to us. The role that emotions play in our decision making was noted earlier by neurologist Antonio Damasio, a researcher at University of Southern California, who analyzed victims that suffered from brain injuries. Dr. Antonio Damasio, first, discovered that a patient, who suffered from ventromedial frontal lobe damage that was caused by a tumor, was incapable of making decisions.

Damasio's research lead him to hypothesis that it was our emotions that guided our behavior not our brain (https://en.m.wikipedia.org/wiki/Somatic_marker_hypothesis, 2018). Damasio, defined emotions as changes in both **the body and the brain** in response to stimuli. He stated that bodily changes are relayed to the brain *where they are transformed into an **emotion*** that tells the individual something about the stimulus. The emotion and the pairing of similar situations guide current behavioral decisions (Damasio, 1994). Damasio (1994), concluded that "Nature appears to have built the **apparatus of rationality** not just on top of the **apparatus of biological regulation**, but also *from* it and *with* it." (p.128.). Damasio's conclusion corrected an earlier assumption by Des Cartes, the philosopher, that our brain alone was responsible for our executive decisions. Damasio emphasized the important role that our body plays in making behavioral decisions. In the next section, Damasio lays the foundation for his reasoning by discussing: a) how some organisms were capable of feelings prior to the brain's development as we know it, b) the difference between stimulus response learning and thought, c) the role of somatic markers, and d) the difference between emotions and feelings.

A. The Insect and Feelings:

In clarifying the role that the body and our emotions have with respect to our behavior, Damasio used the example of an insect's intestinal brain which was used to record the organism's feelings.

> Damasio stated the central nervous system receives (signals from the body), from the bloodstream, the peripheral branches of the nervous system, and the system of intestines called the "enteric nervous system". "The intestine nervous system is actually the first brain….its where the nervous system begin."(Damasio, 1994, p.127). Within the head brain itself, as opposed to the gut brain or the heart brain, the interconnected neural system around these organs- the deepest and most evolutionarily oldest part, **the brain stem**, receives the first input from these bodily signals. As Damasio noted, clusters of neurons in the brain stem, called nuclei, "provide the first full organism integration of body states available to the central nervous system. *These brain stem nuclei exist even in insects- meaning, feelings have been a part of the life of living organisms for hundreds of millions of years. A feeling, then, is essentially some kind of representation of the (states of the body).* As mammals, we have an extensively developed set of regions above the brain stem, giving us a more complex neural passage of signals then an insect (Seigle, 2018, p.127).

B. Stimulus Response Mechanism versus Thought and Critical Thinking:

In addition to the connection between feelings and an organism, Dasmasio also stated that our nervous system has a second function which was to classify things or events as "good" or "bad". In order to do this, Damasio opined that the organism(s) had (developed) a set of preferences-or criteria, biases, or **values to classify threats**. Under their influence and the agency of experience, the repertoire of things categorized as good or bad grows rapidly, and the ability to detect new good and bad things grows exponentially (D'Masio,1994, p.117). In general, Damasio implied that we developed our own criteria for determining what is threatening and non-threatening based on our experiences. In terms of MacLean's brain model, it appears that emotions/feelings were part of the mechanism used to assign values to an events to help us classifying threats.

To illustrate the set of preferences or criteria which differentiates between threatening and non-threatening events, Damasio used a stimulus response mechanism that was associated with Skinner's classical condition learning. Learning that takes place using the stimulus –response mechanism is based on the pleasure –pain principle. Those stimuli that resulted in a pleasurable response were repeated and those stimuli that resulted in a painful experience were to be avoided. Damasio pointed out that adding more neurons between the stimulus and the response didn't make the brain react differently from any other part of the body. Rather, he stated that "as organisms acquired greater complexity, 'brain-caused' actions required more intermediate processing. Other neurons were interpolated between the stimulus neuron and the response neuron, and varied parallel circuits were thus set up, *but it did not follow that the organisms with that more complicated brain necessarily had a mind.* For Damasio, *having (more neurons) between the stimulus and response brain connections, did not make the brain any different from any other organism in the body* (1994).

In order to distinguish the stimulus –response behavior from thought driven behavior, Damasio (1994), stated "Brains can have many intervening steps in the circuits mediating between stimulus and response, and still have no mind, if they do not meet an essential condition: **the ability to display images internally and to order those images in a process called thought. (p89)**. In clarifying this point, Damasio said *thought was the ability to construct images of events that one has experienced and to recall these images.* In general terms, thought is distinguished from our stimulus response actions based on our **self-awareness of the event**, the ability to construct an image of the event, and our ability to retrieve the event with our memory. Based on Damasio definition of thought, I believe thought can take the form of reflection, self-awareness, critical thinking, introspection, or meditation. The ability to be aware of one's behavior through recall is critical to learning and to change. Without this ability to recall images of our own experiences and be aware of our ensuring behavior, we would also *devoid ourselves of the emotions and feelings* that normally accompany our experiences. In many law enforcement situations, a reflex response is not enough. Officers must continuously employ critical thinking to determine what actions optimally fit what circumstances. Neuroscience knowledge can be used as an educational and assessment tool. In the next section, Damasio described the process of how somatic markers facilitate thought.

C. Somatic Markers:

For Damasio, *the mind was the product of internal images which when reflected upon represent thought.* In this section, Damasio explained how our vision is used to collect visual data from the environment and sends this visual data to the brain. Depending on how we subjectively encode our visual cues, our brain will project those experiences upon recall, as thoughts. Damasio stated that these visual cues representing the images were the thoughts that **originated from our bodily sensory inputs.** He stated, *"Images are based directly on those neural representations, and only on those, which are organized topographically and which occur in early sensory cortices."*(Damasio, 1994, p.98.). In vision, these are the **"retinotopic maps"** that **preserve** features of the geometry of the retinal image, and more generally the *"body image" patterns of stimulation that similarly **preserve a map of the whole body**.* These neural representations exploit the geometry of the map to encourage the evocation of the patterns (images) that matter, *when* they matter, and *because* they matter to the body (DaMasio, p.98), as cited in Dennett (1995). In other words, when we interact with our environment, our body records a neurological pattern or neurological picture of how a particular stimuli affected us which results into somatic markers of various strengths depending on how that stimuli (event, person, concept, etc.) affected us. Non-threatening stimuli may be placed in our short term memory; whereas, an intense emotionally traumatizing event may be hard wired in our long term memory. As noted in a previous section, our visual sensor lens works in tandem with our amygdala (emotional brain) which prioritizes what is threatening or non-threatening.

Damasio further stated that, not all exploitation of such **map-like representations amounts to thought**: "In other words, if our brains would simply generate fine topographically organized representations and **do nothing else with those representations**, I doubt we would ever be **(conscious)** of them as images." **What would be missing is "the neural basis for the self" with which these images must be "correlated."**(Damasio, 1994, p.99). Therefore, in order for thought to exist, there has to be a connection between the visual representations and the **(self-awareness)** of these images by the individual.

While all human behavior is learned, not all behavior is conscious to us. In fact, ninety percent of our behavior may be unconscious to us. According to Siegel (2018), our brain is capable of processing information without us being conscious of it. However, Siegel (2018) stated that we are conscious of our thoughts when we can manipulate variables in our working memory (p.55). In an attempt to explain consciousness, Kaku (2011) theorized that consciousness includes at least three basic components: a) sensing and recognizing the environment, b) *self-awareness*, and c) planning for the future by setting goals and plotting strategies. Sensing meant receiving feedback from one's environment. Instead of self-awareness, Siegel (2018) used the term *receptive awareness* which meant being aware of what happening as it is happening (p.49). Lastly, Kaku (2011) described future planning as the ability to anticipate unforeseen consequences and to develop strategies.

Becoming conscious of our somatic markers are important with respect to how we understand our behavioral choices and our wellness concerns. Our somatic markers are representative of our body's accumulated experiences which includes our emotional states that color and influence everything (Damasio, p.198), as cited in Dennett (1995). Damasio theorized that somatic markers were the dominant influence with respect to our behavior not the brain as we thought. In next section, the author distinguishes between two terms that are often used interchangeably but are quite different.

D. Distinguishing Between Emotions and Feelings

In making the distinction between emotions and feelings, D'Amasio (1994) explained that *feelings are the mental experiences* of body states, which arise as *the brain interprets emotions,* where *(emotions) themselves are the physical states* arising from the body's responses to external stimuli. Kitchen (2015) stated that our emotions are instinctual, the *meanings* that they take on and the feelings they prompt, are individually *based on our programming* or our learning of past and present (experiences). To illustrate the difference between emotions and feelings, Kitchen (2015) used the following order of events: I am threatened, I experience fear, and I **feel** horror. Feelings are the aftermath of emotions and are colored by the thoughts, memories, and images that have become linked with that particular emotion. The italicized phrase "based on our programming" emphasizes that we each interpret our emotions and our feelings differently based on our own unique set of experiences.

In law enforcement training, this may be problematic for several reasons: a) subjectively, each individual may interpret and process the environment differently, b) institutionally, there is a concern with respect to whether general training addresses multiple perspectives or the cultural perspectives of most officers, and c) the what types of data are being collected (quantitative/ procedural or qualitative/subjective assumptions). In most officer involved shootings, when asked to explain what happen, most officers respond that at the time of the shooting, they were in *fear of their life*. While this is a correct statement as to what the body was experiencing, there is a need to understand the officer's mental thoughts before and after the event. What was the officer's attention guided to or what was the officer's attention drawn to before the threatening situation. These questions can provide information regarding the contextual framework of an officer involved shootings. To minimize an officer's over indulgence in prejudging situations, some military personnel, such as the Green Beret, practice a technique called the "empty mind" which allows them to suppress their emotions.

Cognizant of the importance role that our emotions play in all behavior, the information provided is offered for the purpose of enabling officers and citizens with the knowledge to help them better management their emotions and feeling when confronted with social conflict situations. Damasio (1994) suggested, **"we need our feelings to guide behavior in an organized manner"**, as cited in (Seigel, 2018, p.133). Most police de-escalation training place an emphasis on written procedures which are left brain representations. This author believes that the emotional side of our decision making is equally important and therefore warrants the same or more time. In the next section, the author identifies several emotional theories that help us to better understand how our emotions and feelings can impact our decision making.

Emotional Theories (8)

According to (Kaku, 2011), emotions are our natural defense mechanism. Emotions tell us what is good for us and what is harmful. The vast majority of things in the world are either harmful or not very useful. When we experience the emotion of "like," we are learning to identify the tiny fraction of things in the environment that are beneficial to us. In fact, each of our emotions (hate, jealousy, fear, love, etc.) evolved over millions of years to protect us from the dangers of a hostile world and help us to reproduce.

Albert Ellis (1991a) developed an emotional model similar to the model cited by Kitchen (2015): I am threatened, I experience fear, and I **feel** horror. The Ellis model which was referred to as the A-B-C Theory of Emotions consisted of the letter A which represented the **activating event** that triggers the emotion. The letter B which represented the **individual's belief system** about the event, and C represents the **consequences or behavior** that the individual decides should be paired with their belief system and the event. In the Ellis model, the individual's belief system is used as a filter to interpret or encode the stimuli. Our belief systems are based on our own subjective experiences. As such, we have to be cautious because our belief systems can represent a **cognitive bias** which may **distort neutral information**. One example of how our belief system can distort neutral information is the self-fulfilling prophesy where we intentionally behave in a manner to influence the outcomes that we expect. A cognitive bias of this nature can be magnified under stressful situations and can result into deadly consequences.

In another research project on emotions and feelings, Schwartz and Clore (1983) found that people can habitually rely on **their present feelings** to make complex judgements as long as their feelings are perceived to be relevant to what or who they are judging. They rather go with their **gut instinct** rather than delay a decision and support it with timely computations (Schwartz and Clore, 1983). Making decisions based on one's gut instinct can be problematic for the reasons just cited. Police shootings occur at a time where emotions are at a peak. In preventing police shootings, research and police training on cover and concealment procedures can be quite beneficial in terms of delaying decisions until more information can be collected.

In another research study, Smith and Ellsworth (1985) indicated that emotions are not just polar opposites (happy v. sad). Their research indicated that each emotion has various properties. Based on their cognitive-appraisal model, Smith and Ellsworth listed the following range or dimensions that each emotion can be characterized by: certainty, pleasantness, attentional activity, control, anticipated effort, and responsibility. Based on their research, police trainers may consider developing an *emotional continuum* similar to the use of force continuum in order to determine if the emotional and feeling responses are in alignment with the officers' procedural use of force responses during simulation training. Another point made by Smith and Ellsworth (1985) was that emotions are a **goal-directed** and that both judgement and choice are influenced until the **emotion-eliciting problem is resolved**. The research by Smith and Ellsworth touches on several areas that could be used to improve de-escalation models and threat assessment models: a) each emotion has a range of properties, b) emotions can have a predisposition based on past experiences, and c) emotions are goal directed.

In closing, a final point made by Keltner, Ellsworth, and Edwards (1993) was that emotions have a **carry over or transfer effect**. They found that anger can be transferred to other individuals or situations, as cited in Lerner and Keltner (2000). This research is relevant because some officers experience post-traumatic stress disorder(s) over the span of their careers and they carry their emotional trauma to other situations. Because our emotional well beings is inseparable from our cognitive or physical well- being, research can be used to improve de-escalation, crisis management, and wellness practices. In the next section, neuroscience is highlighted in a number of law enforcement applications.

Application: Using Neuroscience in Police Training as An Assessment Tool (2017) (9)

Today, most de-escalation models instruct officers to take a number of sequential use of force steps that are proportionate to an officer's threat perception. Currently, there is no mechanism used to measure an officer's emotional state of mind. Neuroscience can be used to measure thoughts and emotions. In general, Nardi (2013) stated that a **word or picture** could be used to trigger a particular area of the brain that is coordinated to an individual's thought, a decision choice, a physical movement, a facial expression, or an expression of empathy. The technology used to measure these responses include: (positron emission tomography aka PET, fMRI, or EEG). These instruments read the brain's activity which can be mapped and decoded before actual behaviors takes place. By using this technology to pin point brain activity, neuroscience can be used as a proactive assessment tool.

Mind Reading Technology (A)

On the topic of mind reading, Kaku (2011) discussed the different between EEG technology and fMRI technology. Kaku (2011) wrote:

> It's been known since 1875 that the brain is based on electricity moving through its neurons, which generates faint electrical signals that can be measured by placing electrodes around a person's head. By analyzing the electrical impulses picked up by these electrodes, one can record the brain waves. This is called an EEG (electroencephalogram), which can record gross changes in the brain, such as when it is sleeping, and also moods, such as agitation, anger, etc. The output of the EEG can be displayed on a computer screen, which the subject can watch.
>
> But one large disadvantage is that the **EEG cannot localize thoughts** to specific locations of the brain. A much more sensitive method is the fMRI (functional magnetic resonance imaging) scan. EEG and fMRI scans differ in important ways. The EEG scan is a passive device that simply picks up electrical signals from the brain, so we cannot determine very well the location of the source. An fMRI machine uses "echoes" created by radio waves to peer inside living tissue. This allows us to pinpoint the location of the various signals, giving us spectacular 3-D images of inside the brain.
>
> The fMRI scan allows scientists to locate the presence of oxygen contained within hemoglobin in the blood. Since oxygenated hemoglobin contains the energy that fuels cell activity, detecting the flow of this oxygen allows one to trace the flow of thoughts in the brain. In fact, fMRI scans can even detect the motion of thoughts in the living brain to a resolution of .1 millimeter, or smaller than the head of a pin, which corresponds to

perhaps a few thousand neurons. An fMRI can thus give three dimensional pictures of the energy flow inside the thinking brain to astonishing accuracy.

Social Distance Screening (B)

Neuroscience can also be used to measure social distance. We are all familiar with the sentiments of love and hate. Semir Zeki and John Romaya (2008), researchers at University College London , stated while love and hate are opposites, they can both lead to irrational behavior. In an attempt to distinguish between the circuitry of these two sentiments, they first sought to *isolate hate* from related emotions such as fear, threats, and danger. In conducting their research, there found that hate has its own circuitry. The **hate circuitry** included structures in the cortex, sub-cortex, frontal cortex, putamen, and insula. One major difference between the love and hate circuitry was that the research showed that when love was measured a larger portion of the brain was deactivated; whereas, when hate was measured a smaller portion of the frontal cortex was deactivated. The researcher explained this difference by stating *that individuals who practice hate want to exercise judgment against others in a more harmful, injurious, and revengeful way.* As the researchers pointed out, the research finding on the "*hate circuitry*" can be used to address cultural, racial, political, and gender issues. This author believes this neuroscience research can be used in a number of police operations such as cultural diversity, hate crime investigations, and recruiting.

Investigative Tool (C)

From a law enforcement standpoint, Arrigo (2007) described how neuroscience can be used in several investigative applications:

> In one criminal justice application, psychiatrist are using neuroimaging via fMRI to take snapshots or "movies of brain activity". In other words, fMRI provides scanning "near-real time, ultra-high resolution, computer-generated models" of brain behavior and processes. These structural scans of the brain permit the operator of this instrument to observe "a subject's neural response to cognitive or sensory-motor tasks." These tasks are linked to how the brain reacts when presented with a series of questions, and these reactions themselves constitute what is measured. Thus, *the operator of an fMRI device can, quite literally speaking, view how "the subject's brain thinks."* While debatable, neuroscience is being considered in the following areas: interrogating suspects of criminal wrongdoing or extracting information from actual violators of the law, counterterrorism or counter-insurgency surveillance strategies, interrogation efforts of U.S. detainees held in custody either domestically or internationally, and capital punishment determinations for juveniles under the age of 18 convicted of capital crimes. These and other criminal justice applications represent a burgeoning trend in cognitive neuroscience to pinpoint, with increasing levels of scientific and objectivist precision,

how the activities of the brain underscore the thoughts and actions of the individual, making determinations of criminal culpability a matter of understanding the fundamentals laws of human biology.(pp.3-4).

As an investigative tool, Kaku (2011) cited research where neuroscience testing could be used as a form of lie detection. Kaku (2011) explains that telling a lie causes more centers of the brain to light up than telling the truth. Telling a lie implies that you know the truth but are thinking of the lie and its myriad consequences, which requires much more energy than telling the truth. Hence, the fMRI brain scan should be able to detect this extra expenditure of energy.

Administering Third Party Judicial Punishment (D)

Based on the concern for sentencing fairness, neuroscience researchers used FMRI to measure how the brain responds to administrating punishment to offenders. According to Ginther et al (2016), the researcher looked at the two factors that influence punishment: a) the offender's mental state and b) the severity of the harm. By examining specify areas of the brain, researchers found that the amygdala is activated when administering punishment. According to Buckholz et al (2008), the **amygdala** was activated when harm relative to the punishment was being consider but not so much with an individual's culpable state. Another group of researchers, Yu et al (2015) found that the amygdala was also activated when both intentionality and harm were being considered. In a study by Treadway et al., 2014), researchers found that even when gruesome crime scenes revealed that the offender's behavior was accidental; then, the amygdala was suppressed. Treadway et al (2014) suggested that the offender's culpability can also be measured by examining an individual's emotional arousal. In the Ginther et al (2016) research, the researchers identified the following four mental states: a) purposeful mental states (angry), b) reckless mental states (aware of risk), c) negligent mental state (didn't notice), and d) blameless mental state (took all precautions). The purpose of the research was to help Judges better understand how to administer punishment in a fair and impartial manner by using neuroscience as an assessment tool. This research links neuroscience to procedural justice principles.

Shoot-Don't Shoot Simulation Training (E)

In addition to using neuroscience as a procedural justice tool, a group of researchers from the University of Alabama published a research article on using neuroscience in their shoot, don't shoot training exercises as part of their research on de-escalation. In framing their work, Houser, Miller, Fonseca (2017) formulated the following research question: How can we understanding brain activity and identify corresponding regions of the brain to help police officers and first-responders in high-stress situations? Based on this research question, the following study was conducted. Listed below is a description of the program and the study's findings:

Program Description (i)

They program description consisted of participating officers waring a cap outfitted with electrodes and a backpack that houses the mobile EEG amplifier. The EEG amplifier was used to read brain activity while they train in a virtual reality simulator. The researchers used the "shoot, don't shoot" exercise to measure participants brain waves. As a novice research project, there were only three officers who participated. The background of the researchers were: Dr. Rick Houser (counselor education), Dan Fonseca (engineering), and Ryan Cook (clinical mental health counselor). These researchers wanted to study the regions of the brain which are activated during high-threat situations. Some of the preliminary findings are listed below:

Findings (ii)

(1) The "beta" wave was associated with thinking. This area of the brain is located in the right temporal parietal junction and is used for making decisions responding to high threat situations. The right temporal-parietal junction is associated with predicting intentions of others, consistent with a major theory in psychology – 'Theory of Mind.' According to the researchers, "An officer who is able to understand the intentions of others may be more effective in making these high risk decisions and consequently lower the risk of shooting a community member, particularly those who are unarmed or an accidental shooting."

(2) In addition to noting thought processes, the researchers also measured the effectiveness of verbal commands in de-escalating.

(3) While this is new research, the researchers stated that this application of neuroscience to policing is a meaningful way to shape the way that officers interact with members of their community.

(4) Data collection included Simulator Exercises, Verbal Commands, and Interviews.

The researchers used a simulator to track each shot and incorporates the exact location of the shot and its outcome. The supervising officer tracks verbal commands of trainees and can alter each encounter relative to the scenario and protocol. Researchers collect the data and also (interview officers after each session) which helps them refine their research procedures.

(5) What the officer was thinking and what the officer saw (perceptual set).

"We talk a great deal about what we learn each time – what the officer was thinking and what they saw that led to the decision they made," Houser said. "It's very complex, and I have a much greater appreciation and understanding for the difficulty of the job."

(6) Goal: Predict the intentions of others.

Ultimately, the researchers would like to see the officers enhance their ability to understand and predict the intentions of others, based on "Theory of Mind."

At present, the above researchers are collecting and analyzing data and seeking funds to continue their research (Houser, Miller, Fonseca, 2017). As neuroscience research is exploring new ground, there are many unanswered questions with respect to how to construct research designs and how to use neuroscience. However, what we do know is that neuroscience research has been shown to detect an individual's thought pattern, detect a decision choice, detect physical movements, detect facial expressions, and to detect expressions of empathy. Therefore, neuroscience has a promising future in law enforcement. In the next section, we look at braining mapping as that component of neuroscience which allows us to identify the areas of the brain that are stimulated by our thoughts.

Neuroscience and Brain Mapping (10)

In this section on brain mapping, the author list some of the key functional areas of the brain. Based on existing research, the following areas of the brain have been identified. For example, the insula is associated with empathy; the prefrontal cortex is associated with personality expression and social behavior; the cingulate cortex is associated with emotions, learning, and memory; the anterior cingulate cortex is associated with error detection; the nucleus accumbens is associated with activating reward stimulation; the amygdala is associated with emotions and our survival instincts; and the hippocampus is associated with short and long term memory (Berridge,2003). Neuroscientist have also found the areas of the brain associated with stereotypes and prejudices. Stereotypical views are mapped in the following areas: the temporal lobe, the anterior temporal lobe, the interior frontal gyrus, the lateral temporal lobe, and the dorsal medial prefrontal cortex. In the case of prejudices, the following areas are affected, the amygdala, the anterior insula, the striatum, and the orbital frontal cortex. The different in areas between prejudices/biases and stereotypes is similar to the different between what was previously stated about emotions and feelings. Prejudices and biases originate from our emotions and stereotypes originate from our cognitive process (Amodio, 2014).

As stated, neuroscience can be applied in a number of law enforcement applications. This author believes that it should be included in the dialogue with respect to criminal justice reform. In the next section, the author proposes a top down bottom up approach to criminal justice reform by developing macro policies to address systemic conditions and by using neuroscience research as self-regulating tools for both police and citizens to mediate de-escalation and social conflict issues.

PART IV CHANGING THE NARRATIVE AND ALTERNATIVE POLICE TRAINING:

Changing the Narrative (11)

Throughout this manuscript, the author has raised the issue that the environment represents the context within which behavior occurs. If we only focus on the behavior of the individual; then, we are only looking at symptoms rather than root causes. Symptoms are an effect. They inform us that there is a problem but they, themselves, are not the core cause(s) of the problem. For example, the high recidivist rates among ex-offenders represent the symptoms that the current theoretical framework, policy decisions, resource allocation and process strategies may not be working properly. Historically, law enforcement has viewed deviant behavior as a character flaw and used punishment as the deterrent to change behavior. As such, many programs are aimed at transforming how the individual thinks. These programs are quite beneficial to program participants. However, there are also flaws in the system. And when we fail to address the flaws in the system which may resemble the same environment that triggered the initial maladaptive behavior, then, we are embracing a *blind spot* that ultimately creates an incubation system where the **"probability of failure is greater than the probability of success"**. Empirical evidence of these blind spots were illustrated in the previous sections along with the adverse impact that these conditions have on the quality of life experienced by both citizens and police.

Today, there has been a shift to reform legal and procedural processes areas such as bail reform, sentencing reform, marijuana de-criminalization, and mental health diversion programs. Yet, more needs to be done on the preventative side. Changing the narrative may require us to change the theoretical model from the deterrent approach to a **systems approach** since many of the factors impacting crime are macro issues (political, social, educational, labor, etc.) that shape the environment and causes many of the systemic conditions that have been described. If we ignore the environment or the systemic conditions; then, *the flow of deviate behavior will continue and only increase*. Developing strategies to address the flow of deviate behavior requires a preventative approach before individuals become individuals develop criminal records. The e need to replace or supplement the existing deterrent model with a systems model is demonstrated by a recent study by the Pew Research Center. In that study, the Pew Research Center indicated that forty states out of fifty states reflected a huge funding disparity between inmate incarceration cost and student education cost. In most of the participating states, the ratio ranged from 2:1 to 4:1 with respect to actual incarceration expenditures in comparison to education expenditures.

When we see longitudinal patterns of fiscal inequities regarding these institutions across the United States; then, this should raises ethical concerns about **our intentions** and our institutional integrity. Seigel (2018) proposed that in order to change behavior that we need to start by evaluating our intentions rather than just focusing our attention on the problem. Our intentions should represent what our desired outcome or goal should look like. *If the outcome does not align with our desired intentions; then, there is a problem.* Seigel (2018) motto reads, "how our intentions (glows) determines where our attention goes, and how our neural firing flows (positive or negative), and to that degree will determine how our neural and interpersonal connections will grow (or deteriorate)" (p.97). Seigel's focus on intentions parallels a similar concern employed by the law enforcement community, regarding a suspect's criminal culpability. Both neuroscience and the law enforcement community use an individual's intention(s) to measure the degree of an individual's ethics, morals, legality, fairness, integrity, justice, empathy, and or compassion. Weighting in on our intentions is equally important with respect to measuring the integrity of those institutions that administer criminal justice policies, strategies, and programs. Effectiveness must be measured by not only a set of participant statistics but also by changes in the environment that provide resources and positive incentives that discourage deviant behavior from a proactive point of view. The African proverb that it takes an entire community to raise a child includes all of the following stakeholders: law enforcement practitioners, educators, clergy, politicians, academians, economist, labor leaders, physicians, etc. working in concert to create an environment which provides incentives that bring out the best in people. The realization of such a vision begins when we share the same noble intentions. Our intentions should be grounded in the spiritual principle that we develop generations to replace and to exceed us, not worship us.

By focusing on our intentions, we allow ourselves to reset and see individuals as human beings with value and not as objects or things to be *processed only after they are in the criminal justice prison pipeline.* In recapping, we need to focus on the systemic conditions that negatively impact the quality of life issues affecting both police and citizens. We also need to complement existing police training with neuroscience research so that we can provide citizens and police with training and educational material which serves to enhance their understanding of the role that the **emotional – feeling component** plays in our decision making. This shift in narrative is necessary because past practices have not changed many of the quality of life issues and wellness issues that both citizens and police experience. Base on this proposed shift, law enforcement is faced with a new mission: "to protect, to serve, and to restore". The role of restoring includes developing and implementing short and long term strategies to elimination "systemic environmental conditions" from a preventative posture. In the next section, the author proposed adding alternative forms of police training that focus more on emotions –feelings training which are anchored in neuroscience.

Police Training Recommendations (12)

This section aims to supplement existing police training. The institution of policing has always been based on a military structure. Policing, as a para-military organization, has adopted many of the characteristics associated with the military such as the hierarchical structures, command and control, authoritarian, and a closed (communication) system. The command and control philosophy coincides with behavioral training where rewards are associated with desired behavior and punishment is used to deter unaccepted behavior. These rewards and punishment tactics are a form of behavioral modification. The same behavioral modification principles are used curb suspect behaviors. According to McNeil (1982, p.viii), the police para-military culture has created the "warrior mentality" within police cultures, as cited in Birzer (2003). For police, their **concern for officer safety** drives the "warrior" mentality. The warrior mentality is part of an officer's survival instinct and has become **hardwired** into the police culture.

In response to the wave of minority shootings across the national, President Obama law enforcement task force recommended policy initiatives that would support a "guardian concept" of policing over a "warrior concept". These two concepts shouldn't be mutually exclusive. Police embrace a number of different situations that require skill sets that oscillates between cognitive and emotional **performance tools.** Taken at face value, the "warrior mentality" is a form of police cultural bias. As with any kind of bias, Amodio (2014) stated that our prejudices are learned behaviors that drive us to discern an "us" versus "them" mentality. These prejudices are a preference for one's own kind (familiarity, safety, security) and therefore, outsiders are viewed as a threat (Amodio, 2014). For police, their **concern for officer safety** drives both the "us" versus "them" and "warrior" mentalities. Because these two mentalities are linked to an officer's concern for safety; they are difficult to change. Policy initiatives are not enough. Because these mentalities are "hardwired" into the police neural network as a form of self –preservation, additional training is required. Rather than **creating an either or situation**, this author suggest that the "warrior mentality" be integrated with the "guardian mentality" by adding training that reflect the emotional and feeling component of decision making which supports the guardian mentality.

During my dissertation on the police code of silence (Hall, 2014), the same "we" versus "them" mentality emerged as a police cultural bias. The study's findings revealed that the police culture is characterized by a taxonomy of occupational fears that included: **officer safety concerns**, whistle blower/retaliation, double standards in discipline, peer pressure, etc. These fears represented various forms of **emotional stressors** that police experienced on the job. In response to those findings, I recommended future research to examine how police officers make critical decisions under fear and stress based on Gilmartin (1986) research on police hypervigilance, Artwohl and Christensen (1997)'s research on deadly force encounters, and Laur (2002)'s research on the anatomy of fear.

My colleague, Dr. Melanie Smith, and I proposed holistic training that focused on the emotional – feeling component of police behavioral decision making. Periodically, the training is updated. The training listed below aims to *complement* existing police de-escalation and crisis management training. Overall, the purpose of the training proposal is to improve the efficacy of police performance and accountability to the public:

o   **Systems Thinking**

To introduce police to a systems thinking approach to analyzing the interdependent and inter-related macro issues that adversely impact deviant behaviors. This approach provides both citizens and police an opportunity to replace a win-lose mentality with a win-win mentality.

System thinking is a strategy proposal to address the systemic conditions that plague communities of color (Senge,1990). One of the virtues of utilizing a system thinking approach for societal or organizational change is that it focuses on proactive and preventative measures.

o   **Organizational Learning**

To introduce police to the concept of organizational learning as the process of creating, retaining, and transferring knowledge within an organization. Organizational learning emphasizes the importance of detecting policy, program, process, and behavioral errors so that the organization can re-organize itself and adapt to changes in the environment. Organizational learning is based in neuroscience because the human mind is design to detect error, recognize patterns, and re-organize itself.

o   **Building a Cycle of Trust**.

This module examines relationships and behaviors that build trust and those that undermine trust Covey (2006). Building trust is part of neuroscience because it is based on empathy, compassion, social intelligence, emotional intelligence, etc. Empathy is measured by mirror neurons. As a component of emotional intelligence, the empathy quotient test can be used to measure how people feel and how they respond to those feelings.

o   **Emotional Intelligence Virtues of Policing**

This module uses the virtues of policing to explore moral behavior and public trust. Moral behavior and public trust are constructs that are interwoven into the foundation of policing based on Constitutional law. How the police are perceived in society is a determinant of the level of trust and confidence held by the public for police officers. The public trust that is delegated to police officers is essential to the functioning of an orderly society. The erosion of public trust and the disruption of community relations in a society can create dangerous conditions for police officers as they attempt to maintain order in society. Emotional Intelligence emerged from research on the amygdala hijack which lies at the heart of neuroscience research (Goleman, 1994).

o   **Dealing with Bias, Prejudice and Stereotypes**

To introduce police to the concepts of prejudice, bias and stereotypes – what they are, how they differ, and how they are learned. This session will also explore the differences between prejudices and stereotypic views. Stereotypic views and prejudices are distinguished in different regions of the brain. As products of our environments, many of our biases are a result of polarization, marginalization, and socialization. This polarization creates both explicit biases and implicit bias. Neuroscience can be used to detect both stereotypes and biases.

o   **Neuroscience- The Body Under Stress**

To introduce police to how the brain works and how the body is affected under stress, fear, and uncertainty. As the 21st Century emerges around the hectic world of police officers so do a new culmination of emotional hazards and stressors. Unfortunately, most people in leadership and helping positions (i.e. police officers, doctors, teachers, coaches, etc.) lose their effectiveness over time because of the cumulative damage from chronic stress, which can cause hardening of the emotions. Gaines and Jermier (1983) stated that police officers now experience increased and unique job stressors due to constant exposure to society' violence, community pressures, administrative demands, and physical hazards. In addition to dealing with stress, fears, and anxieties, neuroscience via MRI testing (brain mappings) can detect brain activities (thought patterns) regarding racial biases, implicit biases, and stereotypic views. These breakthroughs in neuroscience research increase our efficacy in designing emotional-feeling modules that can complement existing procedural de-escalation training.

o   *Transactional Analysis*

To introduce police to the concept of Transactional Analysis as an intervention tool that can assist individuals in developing better self- awareness, personal growth, interpersonal skills, communication skills, and overall performance. TA offers all individuals a method of understanding, processing, and changing negative thoughts. It provides a means of examining learned behavior from three different personality components. TA also deals with identifying racial biases and stereotypes that individuals learn at home, work, or in society (Harris, 2004). It can also be used to understand childhood emotional trauma and self-esteem.

o   *Resilience/Meditation Training*

In resilience training, traumatic experiences can be re-evaluated by altering the perceived value and meaningfulness of the event. This teaches individuals how to reframe trauma emotional events in order to facilitate their recovery. It emphasizes developing a set of positive core beliefs, maintaining a strong spiritual belief system, facing your fears, and developing a conviction that you will prevail in the end. It also emphasizes creating positive statements to affirm yourself (Charney and Southwick, 2012). In mediation training, officers learn to be more focus and aware of their surroundings. Mediation is helpful because it can build skills in the followings areas: attention, concentration, focus, awareness, and clearing our mind. Many scientific studies support various health benefits through mediation such as reducing inflammation, optimizing cardiovascular, immune, epigenetic, and telomerase functions (Siegel, 2018, p.55).

o   *Procedural Justice and Organizational Trust Building*

To introduce police and community members to trust building by applying the law in a fair and consistent manner. This module explores the importance of understanding the perspectives of both the police and the communities that they serve. It examines how these perspectives influence behavior and how police legitimacy is linked to the public's perception that the police are administering the law fairly. Procedural justice is also grounded in neuroscience by virtual of its focus on the citizen's perception of the officer's intention toward them rather than an officer's knowledge of the job (Lind and Tyler, 1988). Neuroscience is currently being used as an assess tool to measure the fairness of Judges in administering sentencing in court cases.

As mentioned, these training strategies are **not an either or proposition** but are integrated with existing training in order to provide officers and citizens with a full range of cognitive and emotional skills to enable them in becoming better problem solvers and decision makers during social conflict situations. The author believes that the proposed training can be used to facilitate the change toward a "guardian mentality", not as a separate entity but as part of a continuum alongside of the "warrior mentality". Given the quality of life issues that many communities of color face, officers need both the proposed emotional and feelings skill sets to go along with their officer safety tactical skills. According to Southwick and Charney (2012), an individual's coping skills include **different** genetic, biological, psychological, social and spiritual factors for each individual. Since no police training will meet the needs of all officers, it would be prudent to offer officers as many tools as possible in order to address subjective differences. Despite the stress and rigors of police work, most officers do an excellent job. The above training proposal and accompanying skills aim to make the officer's job easier.

Conclusion (13)

Crime is a complex problem that requires complex solutions. The purpose of this article is to inform and provide additional information that can and will improve our efficacy in reforming the criminal justice system. None of the suggestions or recommendations are a panacea in themselves but taken together, this author believes that they will increase our probability of success. Finally, this author offers the disclaimer that there is no fail safe method that will prevent random acts of violence or human accidents. However, the author believes that the recommended changes in the law enforcement narrative which include addressing: a) systemic conditions, b) institutional realignment of resources, and c) alternative police training anchored in neuroscience research are significant in addressing the quality of life and wellness issues discussed.

The proposed changes in the narrative are supported by extant research on the correlation between crime and poverty. The effect of poverty on antisocial development in children is based on them being exposed to multiple stressors and cumulative stressors in their environment. Children who grow up in poverty are more likely to engage in antisocial behavior – such as aggression, rule-breaking behavior, and delinquency – than children who grow up in better off circumstances (see reviews by Bradley & Corwyn, 2002; Brooks-Gunn & Duncan, 1997; McLoyd, 1998), as cited by Russell and Odgers (2016). Of major importance is the work by Tremblay et al (2004), their research showed that childhood poverty could be *used as a **predictor*** for antisocial behavior in multiple points in the life of an individual. Children from low-socioeconomic status (SES) backgrounds are more likely to show chronic aggression during the **first 4 years of life**, and are more likely to engage in serious crime and violence during **adolescence** (Bjerk, 2007; Elliott & Ageton, 1980; Jarjoura, Triplett, & Brinker, 2002), and are more likely to continue their involvement in antisocial behavior **as adults** (Fergusson & Horwood, 2002; Lahey et al., 2006; Odgers et al., 2008), as cited in Russell and Odgers (2016). In terms of the social cost associated with antisocial behavior, the United States spends between $1 to $2 trillion dollars per year, as cited in Russell and Odgers (2016).

In addition to the research and the trillions of dollars linked to the social cost of antisocial behavior, longitudinal studies of crime patterns also are notable. According to Kaeble, D. and Cowhig, M. (2018), the nation's prisons, jails, parole, and probation departments population *increased* from 1,842,100 individuals in 1980 to 6,613, 500 individuals in 2016. As of 2011, there were approximately 2.2 million persons incarcerated in either a state or federal prison (Bureau of Justice Statistics, 2012). Globally, the United States is the "incarceration" leader among all countries. Compared to Rwanda and Russia, the next two countries, the United States incarcerates 670 inmates per 100,000 compare to Rwanda with 434 inmates per 100,000 and Russia with 413 inmates per 100,000 (Walmsley, 2013). Another alarming statistic, sixty percent (60) percent of the total number of prisoners incarcerated are people of color (Carson, E.A., 2018). Monetarily, the United States spends $80 billion annual on supervision, confinement, and rehabilitation (U.S. Department of Justice, 2013a). In a more recent update, the Equal Justice Initiative (2017) adjusted the $80 billion by *$100 billion more annually* when adding various police cost, bail bond cost, court cost, and family support cost to cover the expenses associated with Mass Incarceration. The above research should signal that the "disproportionate number of police shootings" are merely the tip of the iceberg and larger problems loom below.

Due to time constraints and the complexity of the issues below the iceberg, little attention is given to root causes for crime. This lack of attention has grown into a "deferred maintenance" of resource inequities affecting many communities of color. This author has suggested a top down approach to address the more complex long term criminal justice strategies. This approach encompasses using a system thinking to reduce deviant behaviors. The system thinking approach focuses on the social, political, economic, educational, and labor factors that contribute to the flow of individuals to the school to prison pipe line. The author coins this flow process as "supply side" policing. In addressing this problem, system thinking can be used to detect not only crime patterns, but multiple causation factors that contribute to the problem. System thinking seeks to assess how these factors can be leveraged to eliminate or reduce the problem. Finally, the system thinking approach seeks to design, coordinate and synchronize these multiple factors into short term and long term plans.

In tandem with this top down approach, the author has suggested a bottom up approach which focuses on the individual in the form of the holistic training for both police and citizens. By focusing on training, the act of learning new information can change the neural circuitry of the individual and lead to a change of their thought process and their behavior. The goal of the proposed training is to improve police and citizens relationships based on each other's concern for performance during stressful encounters. The training will provide police and citizens a better understanding of their emotional and feelings so that they will have the skills to self-regulate themselves. By empowering police and citizens with the capacity to self-regulate themselves, the training has the potential to reduce the number of calls for service that police have to respond to. Therefore, this training has the capacity to reduce the demand for police services.

This author believes that by integrating demand side (bottom up) and supply side (top down) strategies; then, law enforcement can leverage real change. This author also believes that choosing demand side strategies over supply side strategies or vice versa will result in less than optimal outcomes. Therefore, these strategies should be viewed as parts of the same goal. Senge et al (2004) offers an explanation of how problems can continuously resurface or reoccur when we take an either or approach rather than a systems approach. According to Senge, Scharmer, Jaworski, and Flowers (2004), problems repeat themselves because we engage in "reactive learning" where we download, and **act** on old habitual ways of thinking. In addition to Senge et al (2004) explanation, this author believes organizations allow problems to repeat themselves due to limited resources, poor planning, and the absence of a long term "shared" vision. With respect to reoccurring events, Senge et al (2004) stated:

> First, we see ourselves as having had no hand in creating the circumstances. Reactive learning is governed by "downloading" habitual ways of thinking, of continuing to see the world within the familiar categories we're comfortable with. We discount interpretations and options for action that are different from those we know and trust. *We act to defend our interests.* In reactive learning, our actions are actually reenacted habits, and we invariably end up reinforcing pre-established mental models. Regardless of the outcome, we end up being "right". At best, we get better at what we have always done. *We remain secure in the cocoon of own worldview, isolated from the larger world* (pp.10-11).

These habitual ways of thinking may work when past experiences are a good predictor of the future. *However, using past experiences as predictors of the future becomes problematic when past experiences fail to address new variables and new situations.* To avoid the pitfalls of "reactive learning", the above authors state that managers and leaders need to suspend their judgment. This author adds that managers and leaders need to *challenge their existing assumptions* about how things work based on their results. Senge et al (2004) suggested an organizational change model, referred to as Theory U, to address "reactive learning".

Senge et al (2004) described Theory U as a three step change model that including sensing, presencing, and realizing. S*ensing* consisted of observing and connecting with the source of the problem rather than the symptoms. Presencing consisted of retreating and reflecting on the problem or collecting data on the problem. Finally, ***realizing*** consisted of analyzing multiple scenarios and then acting swiftly to resolve the problem. In the *sensing stage*, we begin to suspend our habitual thoughts about the problem. In *presencing stage*, we seek to understand the interdependent parts and how they connect as a whole system. In *realizing stage*, we also, view ourselves not as an observer but as ***co-creators (stakeholders),*** who seek to create a new reality that serves the best interest of humanity (pp.88-92).

Changing our thinking from focusing on the flaws of the individual to include the flaws of the system requires new skills and constant practice. This is why our ethics plays a critical role in deciding what to do when a policy, plan, program, strategy or process is not working. Seigel (2018) uses our intentions as a metric for assessing our personal and institutional integrity with respect to what we do before and what we do after our desired outcomes are achieved or not achieved. Seigel (2018) illustrates this concern by the following motto: "how our intentions (glows) determines where our attention goes, and how our neural firing flows (positive or negative), and to that degree will determine how our neural and interpersonal connections will grow (or deteriorate)" (p.97). By focusing on our intentions or our **ethical concerns**, our intentions should have a cascading effect on our assumptions, the theoretical framework, the structure, the content, the evaluation, and all subsequent strategies that we chose in resolving a problem. The bottom up strategy for change emphasizes the importance of changing our individual thought process via training but more important focusing on our ethics as the key to influencing all other behavioral decisions.

In the past, our level of analysis has always been on the individual not the environment. In most cases, we respond to *what* an individual did. However, to truly understand *why* an individual acted a certain way, we must examine the context within which the behavior occurred. Therefore, by shifting our intentions from only the flaws of the individual to addressing the environment, we can began to fix those inequities in the environment. Some of the quality of life issues cited by the Pew Research Center have existed for decades. **Secondly**, being mindful of the difficult job that police have to perform and their existing training, there is always room for improvement. To that extent, this author has proposed additional training that emphasizes the emotions and feeling component of behavioral decision making which includes neuroscience. Neuroscience has been used in the medical profession for years. Though nascent research, neuroscience is being used and can be used in a number of law enforcement functions: de-escalation, recruitment, procedural justice, leadership, wellness, anger management, cultural diversity, and investigations. This training can benefit both police and citizens by providing them with self- regulating tools to detect, monitor, and manage situations where there is a potential for social conflict. The benefit of the top down and bottom up model is that it is proactive and preventative in scope. It allows both citizens and police to learn in a non-confrontational situation or environment. These strategies can be used as an educational and community outreach platform where the emphasis is on *prevention*.

In concluding, **our awareness** of our environment has always determined who will survive or who will thrive. Understanding the dynamics of how individuals interact with their environment can help us make better decisions involving our behavior. As mentioned, many of the environmental issues cited by the Pew Research Center involve disciplines that extent beyond the traditional boundaries controlled or managed by the police. Therefore, this article is an attempt to inform and suggest that police integrate both the "supply side" of policing with traditional demand side of policing. We cannot continue to separate: the individual's behavior from environmental conditions, the cognitive brain from the emotional brain, human capital from social capital, symptoms from root causes, short term solutions from long term solutions, the warrior mentality from the guardian mentality, and the parts from the whole, and still expect *effective* problem solving. The wellness state of our officers and the quality of life issues experienced by our communities of color should be the **litmus test** for whether or not we are effective in our criminal justice reform efforts. A better understanding of our environment, which this author refers to as "*environmental intelligence*" will determine today, as it did billions of years earlier, whether or not we will survive or thrive. We still have a choice, as well as the capacity, to balance the chemistry of these mutually interdependent parts. But why must we wait..........

Epilogue (14)

When Dr. King, published, *Why Must We Wait, in 1964 he shared many famous quotes. Three quotes are listed below:*

In the face of social stagnation and social change:

Dr. King quoted: "Time itself is neutral; it can be used either destructively or constructively. More and more I feel that the people of ill will have used time much more effectively than have the people of good will. We will have to repent in this generation not merely for the hateful words and actions of the bad people but for the appalling silence of the good people. Human progress never rolls in on wheels of inevitability; it comes through the tireless efforts of men willing to work to be co-workers with God, and without this hard work, time itself becomes an ally of the forces of social stagnation. We must use time creatively, in the knowledge that the time is always ripe to do right."

Dr. King further quoted, "The conservatives who say, "Let us not move so fast," and the extremists who say, "Let us go out and whip the world, "would tell you that they are as far apart as the poles. But there is a striking parallel: They accomplish nothing; for they do not reach the people who have a crying need to be free."

Dr. King also quoted, "The ultimate test of a man or women is not where they stand on positions of comfort and convenience, but where he or she stands on positions of challenge and controversy.

I speak to the needs of those who voices may have been drowned and give praise to the Lord for doing a new thing.

References:

Amodio, D.M. (2014). The Neuroscience of Prejudice and Stereotypes. Retrieved on 7-31-18
    from www.ncbi.nlm.nih.gov/m/pubmed/25186236/. *Nat. Rev Neurosci.*2014 Oct., 15
    (10): 670-82, doi:10.1038/nrn 3800.Epub 2014 Sept. 4.

Anderson, D. A. (1999). The aggregate burden of crime. *Journal of Law & Economics*, 42, 611
    642. doi: 10.1086/467436

Artwohl, A. & Christensen. L. (1997). *Deadly Force Encounters: What a Cops Need to Know to*
    Mentally and Physically Prepare for and Win a Gunfight. Boulder, CO: Paladin Press.

Arrigo, B. A. (2007). Punishment, Freedom, and the Culture of Control: The Case of Brain
    Imaging and the Law. *American Journal of Law and Medicine*;
    Boston Vol. 33, Iss. 2/3, (2007): 457-82.

Aventi, T. J. (2003, August). Officer-Involved-Shootings: What we didn't know has hurt us. The
    Police Policy Studies Council. *Law & Order* magazine August edition. Retrieved on 9-
    11-18 at http://www.theppsc.org/Staff_Views/Aveni/OIS.pdf

Bergland, C. (2013). Cortisol: Why the "Stress Hormone" Is Public Enemy No. 1. Retrieve on
    8-30-18 from www.psychologytoday.com/us/blog/the-athletes-way/...

Bergland, C. (2015, June 17). How Do Various Cortisol Levels Impact Cognitive Functioning ...
    Retrieved on 9-11-18 from https://www.psychologytoday.com/.../how-do-various-
    cortisol-levels-impact-cognitive.

Burridge, K. C. (2003). Pleasures of the brain. *Brain Cogn.* 52, 106–128. doi: 10.1016/S0278-
    2626(03)00014-9

Birzer. M.L. (2003) "The theory of andragogy applied to police training", *Policing: An International Journal of Police Strategies & Management*, Vol. 26 Issue: 1, pp.29-42, https://doi.org/10.1108/13639510310460288

Bjerk, D. J. (2007). Measuring the relationship between youth criminal participation and household economic resources. *Journal of Quantitative Criminology*, 23, 23-39. doi: 10.1007/s10940-006-9017-8

Boyd, L. (2006, November). After watching this, your brain will not be the same | Lara Boyd | TEDxVancouver . Retrieved on 8-30-18 from https://www.youtube.com/watch?v=LNHBMFCzznE

Bradley, R. H., Corwyn, R. F., McAdoo, H. P., & Coll, C. G. (2001). The home environments of children in the United States part I: Variations by age, ethnicity, and poverty status. *Child Development*, 72, 1844-1867. doi: 10.1111/1467-8624.t01-1-00382

Brooks-Gunn, J., & Duncan, G. J. (1997). The effects of poverty on children. *The Future of Children*, 7, 55-71.

Buckholtz JW, Asplund CL, Dux PE, Zald DH, Gore JC, Jones OD, Marois R (2008) The neural correlates of third-party punishment. *Neuron* 60: 930–940. CrossRef Medline

Bureau of Justice Statistics (BJS). 2012, November. "Correctional Population in the United States, 2011." Bureau of Justice Statistics, Office of Justice Programs, U.S. Department of Justice, Washington, DC. http://www.bjs.gov/content/pub/ pdf/cpus11.pdf.

Burt, R.S. (1992) *Structural Holes: The Social Structure of Competition*. Harvard University Press, Cambridge, MA.

Carson, E.A. (2018). *Prisoners in 2016*. Washington, DC: Bureau of Justice Statistics.

Charlesworth E. Nathan RG. *Stress Management: A Comprehensive Guide to Wellness*. New
York. Atheneum: 1984.

Charney, D.S. and Southwick, S.M. (2012). *Resilience The Science of Mastering Life's
Greatest Challenges*, Cambridge University Press. 978-0-521-19563-8

Covey, S.M.R. (2006). *The Speed of Trust*. Free Press, New York, NY.

Chamine, S., (2015). *Positive Intelligence*, Greenleaf Book Group Press, Austin, Texas.

Colucci, A.L., McCleary, J.P., and Ng, Y.J. (2014). *Mathematical Methods in the Social
Sciences. Houston, Texas Police Department Project on Officer-Involved Shooting*.
Northwestern University Weinberg College of Arts and Sciences Mathematical
Methods in the Social Sciences Evanston, Illinois

Data works (2014). A Primer of How Students (And All Humans) Learn. Retrieved on 1-12-19
From https://dataworks-ed.com/blog/2014/07/The-information-processing-model/

Damasio, A.R. (1994). *Descartes' error: emotion, reason, and the human brain*. New York:
Putnam.

Dennett, D.C. (1995). Review Of Antonio R. Damasio, *Descartes' Error: Emotion,
Reason, and the Human Brain*, 1994 in the Times Literary Supplement, August 25, 1995,
pp. 3-4.

Downes, D. and Hansen, K. (2005). *Welfare and Punishment in Comparative Perspective.*

Retrieved on 10-14-18 from

%20Welfare%20and%20Punishment%20in%20Comparative%20Context.pdf
Dept. of Social Policy, and Mannheim Centre for Criminology and Criminal Justice,

London School of Economics. **Dept. of Social Policy, London School of Economics,

and Centre for Longitudinal Studies, Institute of Education, University of London.

Ellis, A. (1991a). The ABC 's of Rational Emotional Theory, *The Humanist*, 51(1), 19-49

Elliott, D. S., & Ageton, S. S. (1980). Reconciling race and class-differences in self-reported and

official estimates of delinquency. *American Sociological Review*, 45, 95-110. doi:

10.2307/2095245

Epstein, S. (1994). Integration of the cognitive and the psychodynamic unconscious. *American*

Psychologist, 49, 709-724.

Equal Justice Initiative (February 6, 2017). Mass Incarceration Cost $182 Billion Every Year

without adding much to public safety. Retrieved on January 8, 2019 from https://eji.org/

News/Mass-Incarceration-Cost-182-Billion-Annually.

Fergusson, D. M., & Horwood, L. J. (2002). Male and female offending trajectories.

Development and Psychopathology, 14, 159-177.

Friedersdorf, C., Sept. 19, 2014, *Police Have a Much Bigger Domestic Abuse Problem Than the*

NFL Does. Retrieved on 9-1-18 from

https://theatlantic.com/national/archive/2014/09/police-officers-who-hit-their-wives-or-
girlfriends/a/article/380329/

Gaines, J. and Jermier, J. (1983) Emotional Exhaustion in a High Stress Organization. *Academy of Management Journal*, 26, 567-586. http://dx.doi.org/10.2307/255907

Gilmartin, K.M. (1986). *Hypervigilance: A Learned Perceptual Set and Its Consequences on Police Stress*, Psychological Services to Law Enforcement, U.S. Dept. of Justice Federal Bureau of Investigation, Edited by Reese and Goldstein, Washington, D.C. 1986, Library of Congress Number 85-600538

Ginther, M.R., Bonnie, R.J., Hoffman, M.B., Shen, F. X., Simons, K. W., Jones, O. D., & Marois, R. (2016). Parsing the Behavioral and Brain Mechanisms of Third-Party Punishment. *The Journal of Neuroscience*, September 7, 2016. 36(36):9420-9434

Goleman, D. (1998). *Working with Emotional Intelligence*. Bantam Books. New York

Goleman, D. (1994). *Emotional Intelligence*. Bantam Books. New York.

Gould, J. (2003). Tribune Brain Concept. Retrieved on 1-13-19 from http://uwf.edu/gould/tribunebrain.pdf.

Gross, J.J., Sutton, S.K., & Ketelaar, T. (1998). Relations between affect and personality: Support for the affect-level and affective-reactivity views. *Personality and Social Psychology Bulletin*, 24, 279± 288.

Hall, J.E. (2014). *A Leadership Perspective on the Police Code of Silence, Police Community Relations, and Procedural Justice*. Capella University

Harris, T. A. (2004). *I'm OK, You're OK*. Harper Row

Healthdirect (2018, April). *The role of cortisol in the body*. Retrieved on 9-11-18 from healthdirect.gov.au-role-of-cortisol-in-the body.

Heider, F. & Simmel, M. (1944). An Experimental Study of Apparent Behavior. *American Journal of Psychology*. 1944, 57, 243-259.

Houser, R., Miller, D., Fonseca (2017). *Don't Shoot: UA Combines VR, Neuroscience in Police Training*, College of Education, College of Engineering, Engineering, Faculty & Staff, Outreach, Research, Science & Technology, University of Alabama.

Jarjoura, G. R., Triplett, R. A., & Brinker, G. P. (2002). Growing up poor: Examining the link between persistent childhood poverty and delinquency. *Journal of Quantitative Criminology*, 18, 159-187. doi: 10.1023/a:1015206715838

Johnson-Laird, P.N., & Oatley, K. (1992). Basic emotions, rationality, and folk theory. *Cognition and Emotion*, 6, 201± 223

Kaeble, D. and Cowhig, M. (2018). Correctional Populations in the United States, 2016. Washington, DC: Bureau of Justice Statistics. Key Statistics: Total Correctional Population. Washington, DC: Bureau of Justice Statistics.

Kahneman, D. (2011). *Thinking, Fast and Slow*. Farrar, Strauss, Giroux, New York.

King, M. L. Jr. (1964). *Why We Can't Wait*. New York: New American Library (Harper & Row), ISBN 0451527534

Kitchen, D. (January 12, 2015). *What's the Difference Between Feelings and Emotions, The best brain possible*, https://thebestbrainpossible.com

Keltner, D. and Lerner JS, (2000). Beyond valence: Toward a model of emotion-specific influences on judgment and choice. *Cognition and Emotion*. 14:473–493.

Keltner, D., Ellsworth, P.C., & Edwards, K. (1993). Beyond simple pessimism: Effects of sadness and anger on social perception. *Journal of Personality and Social Psychology*, 64, 740± 752

Kolhberg, L. (1981). Essays on Moral Development. Vol. 1: The Philosophy of Moral Development. San Fransciso, CA: Harper & Row. ISBN 0-06-064760-4.

Lahey, B. B., Van Hulle, C. A., Waldman, I. D., Rodgers, J. L., D'Onofrio, B. M., Pedlow, S., Keenan, K. (2006). Testing descriptive hypotheses regarding sex differences in the development of conduct problems and delinquency. *Journal of Abnormal Child Psychology*, 34, 737-755. doi: DOI 10.1007/s10802-006-9064-5

Laur, D. (2002). *The anatomy of fear and how it relates to survival skills training.* Retrieved on 10-14-2013 from http://www.lwcbooks.com/articles/anatomy.html

Leaf , C., (2013). *Switch on Your Brain*, Grand Rapids, MI. Baker Books Publishing

LeDoux, J. (1999). *The Emotional Brain* (Phoenix , an Imprint of The Orion Publishing Group), London, 1999) as cited in Rimmele (2003).

LeDoux, J.E. (1996). *The emotional brain: The mysterious underpinnings of emotional life*. New York: Simon & Schuster

Lind, E.A. & Tyler, T. R. (1988). *The Social Psychology of Procedural Justice*. New York, New York: Plenum.

Lopez, W. (July 11, 2016). *Mass incarceration in America, explained in 22 maps and charts.*, Retrieved on 10-3-17 from https://www.vox.com/2015/7/13/8913297/mass-incarceration-maps-charts.

Lowery, W., (July 11, 2016), The Washington Post (2016), *Aren't more white people than black people killed by police" Yes, but no.*

Ludwig, J. (2006, September 19). *The cost of crime: Understanding the financial and human impact of criminal activity.* Testimony to the U.S. Senate Judiciary Committee. May 3, 2013, from http://www.gpo.gov/fdsys/pkg/CHRG-109shrg42938/html/CHRG 109shrg42938.htm

Ludwig, J. (2010). The costs of crime. *Criminology & Public Policy*, 9, 307-311. doi: 10.1111/j.1745-9133.2010.00628.x

McGill University (2005). *The Evolutionary Layers Of The Human Brain.*
Retrieved on 10-18-18 from
thebrain.mcgill.ca/flash/d/d_05/d_05_cr/d_05_cr_her/d_05_cr_her.html

McGill University (2004). *The Amygdala and Its Allies.*

Retrieved on 10-18-18 from
thebrain.mcgill.ca/flash/d/d_04/d_04_cr/d_04_cr_peu/d_04_cr_peu.html#2

McLoyd, V. C. (1998). Socioeconomic disadvantage and child development. *American Psychologist*, 53, 185-204. doi: 10.1037/0003-066x.53.2.185

McNeill, W.H. (1982), *The Pursuit of Power: Technology, Armed Forces, and Society since AD 1000*, University of Chicago Press, Chicago, Ill.

Menkes, J. (2005). *Executive Intelligence*. Harper Collins Publishers, LLC. New York, New

    York.

Morin, R. and Mercer, A. (2017, February 8). *A closer look at police officers who have fired*

    their weapon on duty. Retrieved on 8-15-18 from www.pewresearch.org/fact-

    tank/2017/02/08/a-closer-look-at-police-officers-who-have-fired-their-weapon-on-duty/

Nardi, D. (2013, Aug. 7). *Neuroscience of Personality Your Brain and Type*. Retrieved on 9-11-

    18 from https://www.opp.com/en/Knowledge-centre/Blog/2013/August/Neuroscience-of-

    Personality-Your-Brain-and-Type

Odgers, C. L., Moffitt, T. E., Broadbent, J. M., Dickson, N., Hancox, R. J., Harrington, H., . . .

    Caspi, A. (2008). Female and male antisocial trajectories: From childhood origins to

    adult outcomes. *Development and Psychopathology*, 20, 673-716. doi:

    10.1017/s0954579408000333

Pew Research Center Social and Demographic Trends (2013). *Chapter 3 Demographic and*

    Economic Data by Race retrieved on 9-11- 18 at

    www.pewsocialtrends.org/2013/08/22/chapter-3-demographic-economic-data-by-race/

Police The Law Enforcement Magazine (2018, July, 10). *NLEOMF: Firearms-Related Duty*

    Deaths Up 24% in 2018. Retrieved on august 5, 2018 from

    www.policemag.com/channel/patrol/news/2018/07/10/nleomf-firearms-related-duty-

    deaths-up-24-percent-in-2018.aspx

Pryor-Brown, L., Cowen, E. L., Hightower, A. D., & Lotyczewski, B. S. (1986). Demographic

differences among children in judging and experiencing specific stressful life events. *The*

Journal of Special Education, 20, 339-346. doi: 10.1177/002246698602000307

Ricciardelli, R, Carleton, R. N., Groll, D., Cramm, H. (2018). Qualitatively Unpacking Canadian

Public Safety Personnel Experiences of Trauma and Their Well-Being. *Canadian Journal*

of Criminology and Criminal Justice. Vol. 60, Iss. 4, October 2018, pp. 566-517.

Rimmele, U.G., Domes, M. K., H. M., *Neuroreport* , A Primer on Emotions and Learning, 14,

2485-2488 (2003).

Roufa, T. (2017). Psychological Tests and Screening for Police Officers, *The Balance.*

Russell, M.A. and Odger, C.L. (2016). *Antisocial Behavior Among Children in Poverty:*

Understanding Environmental Effects in Daily Life. Retrieved on Oct. 6, 2018 from

https://adaptlab.org/wp-content/uploads/2016/.../RussellEnvEffChp5.30.13R_corr.pdf

Savage, M. (May, 22, 2018) The Role of Memory in Learning: Encoding. Retrieved on 1-12-

19 from

https://www.obsidianlearning.com/blog/2018/05/the-role-of-memory-in-learning-

encoding.html

Schneider, T. V. (2017, June 22*). If you want it, you might get it. The Reticular Activating*

System Explained. Retrieved on 9-11-18 from https://medium.com/desk-of-van-

schneider/if-you-want-it-you-might-get-it-the-reticular-activating-sysem-explained-

761b6ac14e53

Senge, Peter. 1990. *The Fifth Discipline: the Art and Practice of the Learning Organization*.

New York: Doubleday.

Senge, P., Scharmer, C.O., Jaworski, J., and Flower, B.S., (2004). *Presence: An Exploration, of Profound Change in People, Organizations, and Society.* Currency Double Day, New York.

Siddle, B.K. (1995). *Sharpening the Warriors Edge: The Psychology and Science of Training.*

Siegel, D.J. (2018). *Awareness: The Science and Practice of Presence,* Tarcher Perigee Book, New York, New York.

Southwick, S.M., and Charney, D.S. (2012*). Resilience: the science of mastering life's greatest challenge.* Cambridge University Press.

Schwarz, N., & Clore, G.L. (1983). Mood, misattribution and judgments of well-being: Informative and directive functions of affective states. *Journal of Personality and Social Psychology,* 45, 513± 523.

Schwarz, N., & Clore, G. (1996). Feelings and phenomenal experiences. In E.T. Higgins & A.W . Kruglanski (Eds.), Social psychology: Handbook of basic principles (pp. 433± 465). New York: Guilford Press. Simon, H.A. (1967). Motivational and emotional controls of cognition. *Psychological Review,* 74, 29± 39.

Smith, C.A., & Ellsworth, P.C. (1985). Patterns of cognitive appraisal in emotion. *Journal of Personality and Social Psychology,* 48, 813± 838

Tate, J. Jenkins, J., and Rich, S. (2018). *National Police Shootings.* Retrieved on January 7, 2018 from https://www.washingtonpost.com/graphics/2018/national/police-shootings-2018/?utm_term=.d3f352b086f1

Thomas, H. (2015, August, 21). *Study of holocaust survivors finds trauma passed on to childrens genes*. Retrieved on August 30, 2017 from https://www.theguardian.com/science/2015/aug/21/study-of-holocaust-survivors-finds-trauma-passed-on-to-childrens-genes.

Thomas, T.S. (March 21, 2018). *PTSD: The Dirty Little Secret of Law Enforcement*. Retrieved on 8-30-18 from https://The Crime Report.org/2018/03/21_ PTSD:The dirty –little-secret-of-law-enforcement

Treadway MT, Buckholtz JW, Martin JW, Jan K, Asplund CL, Ginther MR, Jones OD, Marois R (2014) Corticolimbic gating of emotion-driven punishment. *Nat Neurosci* 17:1270 –1275. CrossRef Medline

Tremblay, R. E., Nagin, D. S., Seguin, J. R., Zoccolillo, M., Zelazo, P. D., Boivin, M., . . . Japel, C. (2004). Physical aggression during early childhood: Trajectories and predictors. *Pediatrics*, 114, e43-e50.

University College London. "Brain's 'Hate Circuit' Identified". ScienceDaily. ScienceDaily, 29 October 2008, Retrieved on 1-10-2019 from www.sciencedaily.com/releases/2008/10/08/10282056558.htm

U.S. Department of Justice (DOJ). 2013a. "Smart on Crime: Reforming the Criminal Justice System for the 21st Century."

Walmsley, Roy. 2013. "World Prison Population List (tenth edition)." International Centre for Prison Studies, University of Essex, UK. http://www.prisonstudies.org/sites/prisonstudies.org/files/resources/downloads/wppl_10. pdf.

Walker, T. (2018, April 19). *First Responder PTSD and Suicides, and the Silence That Surrounds Them*. Retrieved on August 3, 2018 from witnessla.com/first-responder-ptsd-and-suicides-and-the-silence-that-surrounds-them/

Weaver, D. (2015, Aug.20). *Why police officers are 30 times more likely to death from heart attacks*. Retrieved on 9-11-18 at Press of AtlanticCity.com. news/Why-police-officers-are-times-more-likely-to-suffer-a/article-Od924afe-478e-11e5-a717-c7065179ed7c.html

Wikipedia (2018, May 22). *Somatic Markers*. Retrieved on 9-11-18 from (https://en.m.wikipedia.org/wiki/Somatic_marker_hypothesis, 2018).

Yu H, Li J, Zhou X (2015) Neural substrates of intention: consequence integration and its impact on reactive punishment in interpersonal transgression. *J Neurosci* 35:4917– 4925. CrossRef Medline

Zeki, et al. (2008). Neural Correlates of Hate. Plos One, 2008; 3 (10): e3556 DOI: 10. 1371/ Journal.pone.0003556

ABOUT THE AUTHOR

Dr. John "Jay" Hall, is a retired Lieutenant that served on the Houston Police Department for twenty three and half years. Dr. Jay is one of nine children, and the only child who attended a four year college. He grew up in Gary, Ind. – Chicago, Ill., and developed his grit from being one of the "Corner Boys". With the some help from some decent cops, who his saw his potential, he was encouraged to go to college. His academic achievements" B. A. Sociology and Criminal Justice from St. Joseph College; M.A. Public Administration from Indiana University NW; M.S.M. in Management from Houston Baptist University; and a Phd. in Organizational Behavioral and Management, with a specialization in Leadership from Capella University. Dr. Hall dedicates any accomplishments to his father who only obtained a fourth grade education, but honorably served in the Military as did all his uncles and brothers with the exception of one. As a baby bloomer, Dr. Hall's world view is still influenced by the civil rights struggle and social justice. He believes in teaching individuals how to fish rather than just giving them a fish. It's part of his tough love campaign for individual growth and social change. His friends call him Jay or Dr. Jay.

- **Naloxone** has no potential for abuse because it does not activate the opioid receptors. It simply blocks the effects of opioids and has no euphoric or psychoactive effects.
- **Naltrexone** also has no potential for abuse. However, it is used to block the pleasurable effects of opioids and alcohol, making it particularly beneficial in addiction treatment. It can reduce cravings and help maintain abstinence by removing the reinforcing effects of substance use.

Impact on Endorphin System:

Both Naloxone and Naltrexone act as inhibitors of the endorphin system. Endorphins, which are the body's natural opioids, bind to the opioid receptors to reduce pain and induce pleasure. By blocking these receptors, Naloxone and Naltrexone prevent the natural endorphins from producing their effects, as well as the effects of external opioids and alcohol.

While Naloxone's primary role is to reverse opioid overdose in an emergency, Naltrexone plays a longer-term role in managing and mitigating the effects of substance use disorders by preventing the body from experiencing the pleasurable effects of opioids and alcohol. This chronic blockade can be instrumental in helping individuals in recovery stay sober and avoid relapse.

In the next chapters, we will explore how these mechanisms of action translate into therapeutic applications, particularly in pain management, addiction recovery, and the potential psychological effects of endorphin inhibition. Understanding these underlying mechanisms is essential for comprehending how both Naloxone and Naltrexone can be leveraged to treat opioid addiction, alcohol use disorders, and other conditions influenced by the endorphin system.

Chapter 6: Endorphin Inhibition in Pain Management
Understanding Pain and the Role of Endorphins

Pain is a complex physiological and psychological experience that plays a crucial role in the body's defense mechanisms. It serves as a signal that something may be wrong, triggering the body's response to injury or harm. The experience of pain is not just a physical sensation; it is influenced by emotional, cognitive, and environmental factors, making it a multifaceted experience.

Endorphins, the body's natural painkillers, are released in response to pain, helping to alleviate the discomfort associated with physical injury or emotional distress. These neurochemicals act by binding to opioid receptors in the brain and spinal cord, leading to a reduction in pain perception. By inhibiting pain signals at various levels of the central nervous system (CNS), endorphins promote a sense of well-being and help individuals endure pain. In situations of acute injury or chronic pain, the body's endogenous endorphin release is an important physiological response that aids in the management of discomfort.

However, in cases of chronic pain, endorphin levels may not be sufficient to cope with persistent pain signals, which can lead to the use of external opioids or pharmacological interventions to manage the symptoms. This is where opioid antagonists, such as Naloxone and Naltrexone, come into play. While these drugs are often used in addiction and overdose management, they also have applications in pain management, particularly in the modulation of endorphin activity and its role in pain relief.

How Naloxone and Naltrexone Are Used in Pain Management

Although Naloxone and Naltrexone are primarily known for their use in addiction treatment, their role in pain management is significant, especially in cases where opioids are not the preferred treatment due to the risk of dependency or addiction. By inhibiting the effects of endogenous and exogenous opioids, these drugs can help manage pain in different clinical scenarios, from acute to chronic conditions.

1. **Naloxone in Pain Management:** Naloxone is typically used as an emergency medication in the treatment of opioid overdose, but it can also be used in the context of chronic pain management. In certain medical conditions, such as opioid-induced hyperalgesia (a phenomenon where opioids cause an increase in sensitivity to pain), Naloxone can reverse the effects of opioids, helping to reduce pain sensitivity. It can be used in conjunction with low doses of opioids to minimize opioid-induced side effects, such as tolerance and hyperalgesia, without compromising the analgesic effects of the opioid itself.

Naloxone is often administered through combination therapies with opioid medications. The low dose of Naloxone is used to prevent the development of tolerance and mitigate potential side effects while still providing pain relief through the opioid. Naloxone can also be used in patients who are on high doses of opioids and may be experiencing adverse effects, such as sedation or respiratory depression.

In clinical settings, Naloxone is sometimes used in the treatment of **opioid overdose-induced pain**. While Naloxone works to reverse opioid toxicity, it can also alleviate the discomfort caused by opioid-induced respiratory depression and the accompanying distress that patients experience during overdose situations.

2. **Naltrexone in Pain Management:** Naltrexone, like Naloxone, blocks opioid receptors, but its effects are used more strategically in chronic pain management. Naltrexone is often used in cases of **chronic pain** where patients are trying to reduce their reliance on opioids, especially in the context of managing conditions such as fibromyalgia, chronic back pain, and osteoarthritis.

   Unlike Naloxone, which is used more acutely, Naltrexone can be used as part of a long-term treatment plan. It helps reduce pain by blocking opioid receptors and preventing the euphoric or reinforcing effects of opioid medications. One interesting application of Naltrexone in pain management is the use of **low-dose naltrexone (LDN)**. In low doses, Naltrexone has been shown to help modulate the immune system and reduce inflammation, which can be particularly helpful in autoimmune and inflammatory pain conditions.

   For example, in conditions like **multiple sclerosis**, **crohn's disease**, and **irritable bowel syndrome**, where inflammation is a major contributor to pain, Naltrexone can help reduce the inflammatory response and decrease pain perception. Low-dose Naltrexone has been explored in the treatment of **neuropathic pain**, where traditional painkillers often fall short, and in patients with conditions where opioid use is contraindicated or potentially harmful.

Clinical Outcomes in Pain Relief with Endorphin Inhibitors

The clinical outcomes of using Naloxone and Naltrexone in pain relief vary depending on the type of pain, the duration of treatment, and the individual patient's response. Both medications have shown positive outcomes in specific clinical scenarios, particularly when conventional opioid treatments are either insufficient or too risky due to the potential for abuse, addiction, or overdose.

1. **Short-Term Pain Relief:** In emergency or acute pain situations, Naloxone is often used to reverse opioid overdose and prevent complications like respiratory depression. In patients who are already receiving opioids for pain, Naloxone can reverse the overdose without causing withdrawal symptoms or losing the analgesic effects of the opioids, provided it is administered in the right context. This makes Naloxone a valuable tool in preventing overdose deaths while providing immediate relief from opioid toxicity.

2. **Long-Term Pain Management:** Naltrexone, especially in low doses, offers a promising approach for patients dealing with chronic pain conditions. By reducing the need for high doses of opioids, Naltrexone can help manage pain while minimizing the risks of opioid addiction and dependence. It has been especially useful in managing conditions that involve inflammation or immune system dysfunction, where traditional opioids may not be the best treatment option. Studies have shown that Naltrexone can improve quality of life for individuals with **chronic pain**, especially when other treatments have failed. Patients using Naltrexone may experience reduced pain sensitivity, decreased inflammation, and improved mobility without the risk of opioid dependency.

3. **Pain in Addiction Recovery:** For individuals in addiction recovery, pain management becomes a significant challenge. The use of opioid medications for managing pain can put individuals at risk of relapse. In these cases, Naltrexone provides a safer alternative, blocking the effects of any opioids consumed and ensuring that individuals do not experience the reinforcing effects of pain relief through opioids. The use of Naltrexone in recovery-based pain management ensures that individuals can receive appropriate pain relief without jeopardizing their sobriety.

Challenges and Considerations in Using Endorphin Inhibitors for Pain

While Naloxone and Naltrexone offer valuable benefits in pain management, there are some challenges and considerations to keep in mind when using these drugs.

1. **Tolerance and Dependence:** While Naloxone and Naltrexone help to block the effects of opioids, they do not address the psychological aspects of pain. This means that in some cases, individuals may still feel pain despite the medication blocking opioid receptors. This can make it difficult for individuals to fully manage pain without complementary therapies, such as physical therapy, cognitive-behavioral therapy, or other non-opioid pain relief methods.

2. **Side Effects:** As with any medication, Naloxone and Naltrexone come with potential side effects. For example, Naloxone can precipitate opioid withdrawal symptoms if administered to someone who is physically dependent on opioids. Naltrexone, especially at higher doses, can cause nausea, fatigue, and mood swings. These side effects must be carefully monitored in a clinical setting, particularly for patients with chronic pain or those in addiction recovery.

3. **Need for Integrated Care:** Pain management using Naloxone and Naltrexone is most effective when integrated into a comprehensive care plan. These medications should be combined with non-pharmacological treatments for pain, such as therapy, acupuncture, or other alternative treatments, to ensure optimal outcomes.

Conclusion

Naloxone and Naltrexone offer important benefits in pain management, particularly in cases where opioid medications are not ideal or carry significant risks of addiction and overdose. Naloxone's role in managing opioid-induced pain and reversing overdose is invaluable, while Naltrexone's long-term benefits in chronic pain and addiction recovery make it a powerful tool for individuals seeking to manage pain without resorting to opioids. By understanding the role of endorphins in pain and how these inhibitors work to modulate that system, clinicians and patients alike can make more informed decisions about pain treatment strategies in the context of addiction and chronic conditions.

Chapter 7: Psychological Effects of Endorphin Inhibition
The Relationship Between Endorphins and Mood

Endorphins are often referred to as the body's natural mood boosters. As endogenous opioids, they play a crucial role in regulating mood, emotional responses, and stress resilience. When endorphins bind to opioid receptors in the brain, particularly the mu-opioid receptors, they trigger the release of other neurotransmitters such as dopamine and serotonin, which are known for their roles in promoting feelings of pleasure, happiness, and well-being.

Endorphins are released in response to both physical activities—such as exercise, laughter, and sexual activity—and psychological stimuli, like positive social interactions or achieving a sense of accomplishment. These "feel-good" chemicals are essential for helping individuals cope with stress and pain, as they not only dull the perception of physical discomfort but also elevate mood, reduce anxiety, and promote feelings of euphoria.

The psychological effects of endorphins are particularly important when considering how they interact with conditions like depression, anxiety, and stress. Low levels of endorphins can contribute to emotional dysregulation, making individuals more susceptible to feelings of sadness, irritability, or even chronic stress. As such, maintaining a healthy balance of endorphins is crucial for emotional stability and overall mental health.

Impact of Endorphin Inhibition on Anxiety and Depression

Endorphin inhibition, particularly through the use of opioid antagonists like Naloxone and Naltrexone, can have significant effects on mood and mental health, as these medications block endorphin activity at the opioid receptors.

1. **Naloxone and Mental Health:** Naloxone, typically used in opioid overdose reversal, has a rapid onset of action, blocking the effects of opioids on the brain's opioid receptors. In the context of mental health, Naloxone's effects are generally transient, and its use is focused on reversing overdose-induced suppression of breathing rather than managing mood disorders.

   However, Naloxone's effect on endorphins can provide some insight into how blocking opioid receptors might influence mood. While Naloxone is not typically used to treat anxiety or depression, it can cause **withdrawal-like symptoms** in people who are dependent on opioids. For individuals who are undergoing opioid withdrawal, the abrupt blocking of endorphin activity can lead to heightened anxiety, restlessness, and even depressive symptoms. These temporary effects underscore the critical role endorphins play in mood regulation and how their inhibition can contribute to psychological distress.

2. **Naltrexone and Mental Health:** Naltrexone, like Naloxone, is an opioid antagonist, but its long-term use in the treatment of addiction means that its effects on mood and mental health are more pronounced. By blocking endorphins and preventing opioids from producing their rewarding effects, Naltrexone can reduce the reinforcing cycle of addiction and help individuals maintain sobriety.

   However, Naltrexone's inhibition of endorphins can also have an impact on mental health. For individuals struggling with **opioid use disorder (OUD)** or **alcohol use disorder (AUD)**, blocking the opioid receptors prevents the euphoric effects of these substances. This is beneficial in preventing relapse but may also lead to some individuals experiencing **reduced motivation, irritability**, and **decreased emotional responsiveness**, especially in the early stages of treatment. The lack of the natural "reward" from endorphins can create feelings of emotional flatness or apathy in some patients, especially if they have not yet learned how to derive pleasure from other activities.

   Importantly, however, research has shown that these negative mood effects tend to diminish over time, as patients adjust to the medication and experience long-term benefits, including reduced cravings and better emotional regulation.

How Naloxone and Naltrexone Influence Mental Health

Both Naloxone and Naltrexone are opioid antagonists, meaning they block the effects of opioids and endorphins on the brain. Their influence on mental health varies depending on the context of use:

1. **For Addiction Recovery:** In addiction recovery, **Naltrexone** is used to help individuals manage their cravings and reduce the rewarding effects of opioids and alcohol. By preventing the euphoric effects of these substances, Naltrexone helps people stay sober, which is crucial for long-term recovery.

   However, because Naltrexone blocks endorphin activity, it can have a psychological impact, particularly in the early stages of treatment. Some individuals may initially struggle with feelings of depression, irritability, or anxiety as they adjust to the absence of the rewarding effects of opioids or alcohol. These initial effects tend to subside as the patient's brain chemistry adjusts and as the individual learns new coping mechanisms for managing stress and finding pleasure in other aspects of life.

2. **For Pain Management:** Both Naloxone and Naltrexone, in the context of pain management, can influence mood by modulating the endorphin system. Chronic pain is often accompanied by psychological symptoms, including depression and anxiety, due to the ongoing physical discomfort and the toll it takes on mental health. By inhibiting the effects of endorphins, these medications can change how a person responds emotionally to pain.

   For individuals dealing with **opioid-induced hyperalgesia** (increased pain sensitivity due to prolonged opioid use), Naloxone can reduce pain sensitivity and may improve mood as part of the process of decreasing reliance on opioids. In contrast, Naltrexone's role in managing chronic pain conditions may have an indirect effect on mood by helping to reduce the need for opioids, thus decreasing the emotional cycle of pain and relief that can contribute to mood instability.

Impact of Endorphin Inhibition on Stress Resilience

Endorphins are not only involved in regulating pain and mood but also in helping individuals respond to stress. They play a role in modulating the **hypothalamic-pituitary-adrenal (HPA)** axis, which is responsible for the body's stress response. Endorphins help manage the body's reaction to stressors by mitigating the emotional and physical impacts of stress.

When endorphin levels are high, individuals generally feel more resilient in the face of stress and anxiety. However, when endorphin levels are inhibited, as with the use of Naloxone or Naltrexone, individuals may experience reduced stress tolerance and greater vulnerability to stress-induced emotional or physical symptoms. This can manifest as heightened anxiety, irritability, or a reduced ability to cope with life's challenges.

Balancing Endorphin Inhibition for Mental Health Benefits

The key to using Naloxone and Naltrexone effectively in treating addiction and pain while managing their psychological side effects lies in balance. While blocking endorphins can reduce cravings and prevent relapse in addiction, it can also lead to mood disturbances or emotional flatness. As such, treatment should be paired with behavioral therapies that help individuals develop emotional resilience and coping strategies for stress and mood regulation.

Patients using Naltrexone or Naloxone should be closely monitored, especially during the early stages of treatment, to ensure that any emotional or psychological side effects are managed appropriately. Support systems, including counseling, therapy, and peer support, can provide individuals with the tools they need to adjust to the psychological changes associated with endorphin inhibition.

Conclusion

Endorphins are crucial for regulating mood, emotional responses, and stress resilience. The inhibition of endorphins, as achieved by Naloxone and Naltrexone, can have significant psychological effects, including changes in mood, anxiety, and stress tolerance. While these medications are invaluable in addiction treatment and pain management, they must be used carefully, with an understanding of their potential mental health impacts. By combining pharmacological treatments with comprehensive mental health support, patients can navigate the challenges of endorphin inhibition and experience lasting recovery and well-being.

Chapter 8: The Risk and Side Effects of Naloxone
Common Side Effects of Naloxone

Naloxone is a life-saving medication, primarily used in the treatment of opioid overdose, and its rapid action in reversing opioid toxicity can significantly reduce the risk of death in emergency situations. However, like any medication, Naloxone is not without side effects. While it is considered safe and well-tolerated, understanding the potential risks is important for both healthcare providers and patients.

The most common side effects of Naloxone are typically mild and occur due to the rapid reversal of opioid effects in the body. These can include:

Withdrawal Symptoms:

- Agitation

- Nausea and vomiting

- Sweating

- Anxiety

- Tachycardia (rapid heart rate)

- Muscle aches and pains

- Diarrhea These symptoms, while unpleasant, are typically short-lived and subside once the Naloxone wears off. However, they may be distressing for the individual experiencing them, especially if they have a high degree of opioid dependence.

2. **Headache:** Headaches are another common side effect of Naloxone, likely related to the physiological changes that occur when the drug reverses opioid-induced respiratory depression and alters the body's normal state. Headaches usually resolve once the individual stabilizes and the effects of Naloxone wear off.

3. **Dizziness or Lightheadedness:** Because Naloxone works by reversing the effects of opioids, including their impact on the central nervous system, it can sometimes cause dizziness or lightheadedness. This can occur particularly when administered to individuals who have been sedated or are in respiratory distress due to opioid overdose.

4. **Increased Blood Pressure:** Some individuals may experience an increase in blood pressure after receiving Naloxone, especially if they were previously experiencing hypotension (low blood pressure) due to opioid overdose. This effect is typically transient and resolves after the initial effects of Naloxone wear off.

5. **Reversal of Pain Relief:** Naloxone can also reverse the analgesic effects of opioids, which may result in the return of pain in individuals who were previously receiving opioids for pain management. This can be particularly challenging in emergency or chronic pain settings and may require additional pain management strategies after Naloxone administration.

Rare and Severe Reactions to Naloxone

While rare, there are some severe reactions that can occur when administering Naloxone, particularly when it is used inappropriately or in individuals with pre-existing conditions. These include:

1. **Severe Allergic Reactions (Anaphylaxis):** Although exceedingly rare, some individuals may experience an allergic reaction to Naloxone. Symptoms of a severe allergic reaction can include difficulty breathing, swelling of the face or throat, and hives. If any of these symptoms occur, immediate medical attention is required.

2. **Severe Withdrawal Symptoms:** In individuals with a long history of opioid use or those with high opioid tolerance, Naloxone can precipitate **severe withdrawal symptoms**. This may include intense agitation, hallucinations, and severe discomfort. In such cases, individuals may require additional medical monitoring and support to manage the withdrawal process effectively.

3. **Seizures:** In rare cases, the sudden reversal of opioid overdose with Naloxone can result in seizures. This is generally seen in cases of mixed drug overdoses where the individual has used not only opioids but other central nervous system depressants, such as benzodiazepines or alcohol. Seizures can also occur if Naloxone is administered rapidly or in excessive doses.

4. **Cardiovascular Issues:** Rare cardiovascular effects have been reported with Naloxone, especially in individuals with pre-existing heart conditions. This can include arrhythmias (irregular heartbeats) or severe hypertension. Such cases are uncommon and typically occur when Naloxone is administered in high doses or to individuals with underlying cardiovascular disease.

One of the most important characteristics of Naloxone is its rapid onset of action. When administered, Naloxone works quickly to reverse the effects of opioid overdose, particularly respiratory depression, which is the leading cause of death in opioid toxicity. However, Naloxone's relatively short duration of action can present challenges in certain clinical situations.

1. **Short Duration of Effect:** Naloxone's half-life ranges from 30 minutes to 1 hour, which means that its effects can wear off quickly. If the opioid involved in the overdose has a longer half-life, such as **fentanyl** or **methadone**, Naloxone's effects may dissipate before the opioid toxicity is fully reversed. This can lead to a return of overdose symptoms and requires close monitoring and potentially additional doses of Naloxone.

2. **The Need for Repeat Dosing:** In cases of overdose with long-acting opioids or high-potency opioids, the rapid onset and short duration of Naloxone's effects mean that multiple doses may be necessary to completely reverse the overdose. This requires vigilant monitoring and timely re-administration to ensure the individual does not relapse into overdose.

3. **Clinical Management in Acute Settings:** The rapid onset of Naloxone is crucial in emergency medicine, as it allows for immediate reversal of overdose symptoms, potentially saving lives within minutes. However, healthcare providers must be prepared for the potential need for repeat doses and should have plans in place for continued monitoring of the patient. In some cases, once the initial overdose is reversed with Naloxone, individuals may require intensive care or long-term support to manage the underlying cause of the overdose and any other complications that may arise.

Conclusion: Managing the Risks and Side Effects of Naloxone

Naloxone is a powerful tool in the treatment of opioid overdose, and its quick action can save lives. While it is generally safe and well-tolerated, it is important to understand the potential side effects and risks associated with its use. Common side effects like withdrawal symptoms, headaches, and dizziness are typically short-lived, but healthcare providers should be prepared to manage more severe reactions when they arise.

By understanding the pharmacodynamics and pharmacokinetics of Naloxone, healthcare professionals can better anticipate how the drug will act in emergency situations and adjust treatment protocols accordingly. Careful monitoring, repeat dosing when necessary, and comprehensive patient management are key to ensuring that Naloxone is used effectively and safely in reversing opioid overdose and saving lives.

Chapter 9: The Risk and Side Effects of Naltrexone
Common Side Effects of Naltrexone

While Naltrexone is a highly effective medication in the treatment of opioid and alcohol use disorders, like all medications, it comes with potential side effects. It is crucial for healthcare providers and patients to understand both the common and rare side effects to ensure proper use and management of the medication.

Gastrointestinal Symptoms:

- **Nausea:** Nausea is commonly reported by individuals who are starting Naltrexone. This side effect tends to decrease as the body adjusts to the medication.

- **Vomiting:** In some cases, nausea can progress to vomiting, especially in the first few days of treatment.

- **Loss of Appetite:** Naltrexone may cause a decrease in appetite, leading to weight loss in some individuals, which may be a concern if the person is already underweight or malnourished.

2. These side effects are typically temporary and resolve as the individual continues with treatment. Taking the medication with food may help alleviate some of these symptoms.

3. **Headache:** Headaches are another relatively common side effect, particularly during the initiation of treatment. These are often mild to moderate in intensity but can be bothersome. Hydration, adjusting the dose, or managing through over-the-counter pain relief can help in most cases.

4. **Fatigue and Sleep Disturbances:** Naltrexone may cause feelings of fatigue, drowsiness, or insomnia. For some individuals, sleep disturbances can make it difficult to stay on the medication long-term. Managing these side effects may involve adjusting the timing of the medication, such as taking it earlier in the day to avoid interference with sleep at night.

5. **Dizziness:** Some individuals report dizziness or lightheadedness when taking Naltrexone. This side effect is generally mild and transient but should be monitored, especially in individuals who may have low blood pressure or other cardiovascular issues.

6. **Anxiety and Irritability:** Although Naltrexone is used to reduce cravings and withdrawal symptoms in addiction recovery, some individuals may experience an increase in anxiety, irritability, or mood swings. These symptoms can be
Long-Term Side Effects and Considerations particularly noticeable in the early stages of treatment as the body adjusts to the changes in endorphin activity.

While Naltrexone is generally well-tolerated, its long-term use in the management of opioid and alcohol use disorders requires careful monitoring for potential side effects and considerations. These effects may become more apparent after prolonged use.

1. **Liver Function and Hepatotoxicity:** One of the most serious side effects of long-term Naltrexone use is its potential impact on liver function. Naltrexone is metabolized in the liver, and in rare cases, it can cause **hepatotoxicity** (liver damage). Patients with pre-existing liver conditions, such as hepatitis or cirrhosis, should be closely monitored. Liver enzyme tests should be conducted before starting Naltrexone and periodically throughout treatment to ensure that liver function remains within normal limits.

   If liver toxicity is suspected, Naltrexone should be discontinued immediately, and alternative treatments for addiction recovery should be considered.

2. **Psychological and Emotional Effects:** Although Naltrexone helps in reducing cravings and relapse for opioid and alcohol use disorders, it can cause emotional side effects. In some cases, it may contribute to:

- **Depression:** Some individuals report experiencing depressive symptoms during Naltrexone therapy. The mechanism behind this is not entirely understood, but it may be related to the disruption of the endorphin system, which affects mood regulation.

- **Suicidal Thoughts:** Rarely, Naltrexone has been associated with an increase in suicidal thoughts and behaviors, particularly when taken by individuals with a history of mental health conditions. This highlights the importance of screening for psychiatric conditions and close monitoring for individuals with a history of depression or suicidal ideation.

3. These effects warrant careful screening prior to initiating treatment and ongoing mental health monitoring during Naltrexone therapy. In some cases, dose adjustments or combining Naltrexone with antidepressant therapy may be necessary.

4. **Pain Sensitivity:** Naltrexone works by blocking opioid receptors, which also impacts the body's ability to experience pain relief from opioid analgesics. Patients using Naltrexone may experience **increased sensitivity to pain** (hyperalgesia). This is particularly important for individuals with chronic pain who may require non-opioid pain management strategies.

In such cases, it may be necessary to adjust pain management regimens or incorporate other therapeutic modalities, such as physical therapy, cognitive-behavioral therapy for pain, or non-opioid pain relievers.

How Naltrexone Interacts with Other Medications

Because Naltrexone is an opioid antagonist, it has the potential to interact with a variety of medications, particularly those involving the opioid system or affecting the liver's enzyme activity.

1. **Opioid Analgesics:** The most significant interaction is with **opioid medications**. Naltrexone blocks the effects of opioids, meaning that if a person taking Naltrexone uses opioids (either recreationally or for medical purposes), they will not experience the typical pain-relieving effects. In fact, using opioids while on Naltrexone can precipitate **acute withdrawal symptoms**, which can be severe and dangerous.

   Therefore, individuals taking Naltrexone must be opioid-free before starting treatment and should not use opioids during therapy. This is especially important for patients in recovery from opioid addiction, as it can prevent relapse and reduce the risk of overdose.

2. **Alcohol and Medications Affecting the Liver:** Naltrexone is often used in the treatment of **alcohol use disorder** and can interact with medications that affect liver function. Certain medications, particularly those metabolized in the liver, may require dose adjustments when taken alongside Naltrexone to avoid liver toxicity or altered drug efficacy.

   Drugs that may interact with Naltrexone include:

- **Medications for depression or anxiety**: Some antidepressants, like selective serotonin reuptake inhibitors (SSRIs), may be used concurrently with Naltrexone, but careful monitoring is required to avoid side effects such as serotonin syndrome, especially with serotonergic medications.

- **Anticonvulsants**: Medications like phenytoin or carbamazepine, which affect liver enzymes, may interfere with the metabolism of Naltrexone, reducing its effectiveness.

3. As with any medication, it is essential for patients to disclose all current medications to their healthcare providers before starting Naltrexone.

4. **Over-the-Counter Medications:** Naltrexone may also interact with over-the-counter medications, particularly those containing **opioid derivatives** (like some cough suppressants or diarrhea medications). These should be avoided during Naltrexone therapy, as they may induce withdrawal symptoms or reduce the effectiveness of the treatment.

Conclusion

Naltrexone is a valuable tool in the treatment of opioid and alcohol addiction, but its use comes with potential side effects and risks that must be carefully managed. While common side effects such as nausea, headache, and gastrointestinal discomfort are generally short-lived, long-term use requires vigilant monitoring for liver toxicity, depression, and emotional changes. The risk of opioid withdrawal due to its action on opioid receptors is another critical consideration, especially in individuals who are dependent on opioids.

By understanding the side effects, interactions, and appropriate monitoring strategies, healthcare providers can ensure that Naltrexone is used safely and effectively. Additionally, by integrating mental health support and non-opioid pain management strategies, patients can experience better outcomes and a smoother transition through the challenges of addiction recovery.

Chapter 10: Naloxone and Naltrexone in Opioid Addiction Recovery

The Role of Naloxone in Preventing Opioid Overdose Deaths

Opioid use disorder (OUD) has reached epidemic proportions worldwide, and overdose deaths related to opioid use have become a significant public health crisis. Naloxone has proven to be a critical tool in reducing opioid overdose deaths. As an opioid antagonist, Naloxone can rapidly reverse the life-threatening effects of opioid overdose, particularly respiratory depression, which is the leading cause of death in opioid toxicity.

In an overdose situation, opioids bind to receptors in the brain and nervous system, slowing down breathing and heart rate to dangerously low levels. Naloxone works by displacing opioids from the mu-opioid receptors and reversing these effects. It can restore normal breathing and consciousness within minutes, significantly reducing the risk of death in overdose situations.

One of the most important aspects of Naloxone's role in opioid addiction recovery is its potential to save lives in emergency situations. Access to Naloxone for first responders, healthcare professionals, and individuals at risk of overdose (and their families) is a cornerstone of harm reduction strategies. In some areas, Naloxone is distributed as part of public health initiatives aimed at reducing overdose deaths. These programs have shown success in lowering the mortality rates associated with opioid use.

Furthermore, Naloxone is now widely available through programs that allow non-medical individuals to administer it in the event of an overdose. This has empowered communities to take proactive steps in preventing overdose fatalities, particularly in areas with high rates of opioid addiction.

Naloxone in Emergency Medicine: Protocols and Guidelines

In emergency medical settings, Naloxone is administered to individuals who have experienced an opioid overdose. The drug's administration follows established protocols designed to maximize its effectiveness and minimize the risk of harm.

1. **Initial Assessment:** The first step in the management of a suspected opioid overdose is the assessment of the patient's airway, breathing, and circulation (ABCs). If the individual is unresponsive and shows signs of respiratory depression (slow or absent breathing), Naloxone should be administered immediately.

2. **Administration Methods:** Naloxone can be administered intravenously, intramuscularly, or intranasally. The choice of method depends on the circumstances and available resources. Intranasal Naloxone is particularly useful in pre-hospital settings or where intravenous access is difficult. The intramuscular injection is typically used when intravenous access is available.

3. **Initial Dose and Reassessment:** The usual starting dose of Naloxone is 0.4–2 mg, administered either intranasally or via injection. If there is no response after 2–3 minutes, additional doses may be administered. The total dose administered can be adjusted based on the severity of the overdose, but care must be taken not to overshoot the reversal, which may cause withdrawal symptoms.

4. **Monitoring and Follow-Up:** After Naloxone administration, the individual should be monitored closely for the return of overdose symptoms, especially since Naloxone's effects are short-acting compared to some opioids, which may require re-administration. Individuals should be transported to the hospital for further care and observation if necessary.

Emergency protocols should emphasize rapid access to Naloxone, proper training for administration, and the ability to monitor the patient's response to ensure that they are stable and safe. These guidelines are increasingly part of training programs for first responders, healthcare workers, and members of the public.

Naltrexone in Relapse Prevention for Opioid Addiction

While Naloxone plays a pivotal role in preventing opioid overdose deaths, Naltrexone is used in the long-term management of opioid addiction. As an opioid antagonist, Naltrexone helps individuals in recovery maintain abstinence by blocking the euphoric effects of opioids, making it less likely for individuals to relapse.

Naltrexone's primary use is in preventing relapse after detoxification, as it reduces cravings and blocks the reinforcing effects of opioids. The key benefit of Naltrexone in addiction recovery is that it does not produce a high or lead to addiction itself. This makes it a valuable option for individuals seeking to remain sober without the risk of developing a new substance dependency.

1. **Role in Opioid Addiction Recovery:** After detoxification, Naltrexone is typically introduced to help reduce the risk of relapse. By blocking opioid receptors, it ensures that if an individual relapses and uses opioids, they will not experience the usual rewarding effects. This creates an important deterrent against continued opioid use, as the person will not experience the "high" associated with their substance use.

2. **Long-Term Use and Maintenance:** Naltrexone is used as part of a maintenance regimen for individuals in recovery. The medication is typically taken in oral form once a day, or it can be administered as an extended-release injection (Vivitrol) once a month. The extended-release form offers convenience for patients who have difficulty adhering to a daily medication schedule and helps provide a steady level of opioid blockade over time.

3. **Reducing Cravings:** Naltrexone plays a crucial role in reducing cravings for opioids. Cravings are one of the most challenging aspects of opioid addiction recovery, and Naltrexone's ability to block the rewarding effects of opioids helps individuals stay committed to their recovery process. By diminishing the intense urge to use opioids, Naltrexone allows individuals to focus on behavioral therapies and other aspects of their recovery without being overwhelmed by their addiction.

4. **Combining with Behavioral Therapy:** While Naltrexone is highly effective in relapse prevention, it works best when combined with **behavioral interventions**, such as cognitive-behavioral therapy (CBT), contingency management, and support groups like Narcotics Anonymous (NA). Behavioral therapy helps individuals address the underlying psychological and social triggers that contribute to their addiction. When used together, Naltrexone and behavioral therapy offer a comprehensive approach to long-term recovery.

Combining Naloxone and Naltrexone with Behavioral Interventions

The most effective approach to opioid addiction recovery involves combining pharmacological treatments, such as Naloxone and Naltrexone, with behavioral interventions. While Naloxone serves an immediate, life-saving function in emergency situations, Naltrexone provides long-term support to help individuals remain sober after the crisis has passed.

Behavioral Interventions play a key role in addiction recovery by helping individuals:

- Develop coping strategies for managing cravings and stress.
- Address negative thinking patterns that contribute to addiction.
- Build healthy habits and social support systems.
- Learn how to avoid relapse triggers and high-risk situations.

Integrating Naloxone and Naltrexone with these interventions not only addresses the neurochemical aspects of addiction but also provides the psychological tools needed for sustainable recovery.

Conclusion: The Role of Naloxone and Naltrexone in Opioid Addiction Recovery

Naloxone and Naltrexone are two critical medications in the fight against opioid addiction. Naloxone's rapid action in reversing overdose and preventing opioid-related deaths has saved thousands of lives, making it a cornerstone of harm reduction strategies worldwide. Naltrexone, on the other hand, plays a vital role in long-term addiction recovery by blocking the euphoric effects of opioids, reducing cravings, and preventing relapse.

Together, Naloxone and Naltrexone provide a comprehensive approach to managing opioid addiction—both in emergency settings and as part of ongoing recovery. By integrating these medications with behavioral therapies, individuals in recovery can overcome the neurobiological and psychological challenges of addiction, leading to better outcomes and a more sustainable, healthy life.

Chapter 11: The Future of Naloxone and Naltrexone
Research and Clinical Trials Underway

The future of Naloxone and Naltrexone lies not only in their established uses but also in the expansion of their applications through ongoing research and clinical trials. Both of these medications have proven invaluable in the treatment of opioid use disorder, alcohol dependence, and opioid overdose. However, researchers are continually exploring new ways these medications can be used, both in addiction recovery and in other medical conditions.

One area of focus in ongoing research is **combination therapies**. By combining Naloxone or Naltrexone with other medications, researchers aim to enhance their effectiveness in treating addiction and managing pain. For example, combining Naltrexone with **cognitive-behavioral therapy** (CBT) has shown promise in reducing cravings and preventing relapse, leading to better long-term recovery outcomes. Ongoing studies are also exploring **genetic factors** that may influence the effectiveness of these medications, helping to identify patients who are more likely to benefit from specific treatments.

Naloxone, which is primarily used in emergency overdose situations, is also being studied for its potential to be used in new settings, such as in **preventative measures for those at high risk of overdose**, or as a treatment for other types of substance use disorders, including methamphetamine or cocaine addiction. These studies could result in expanded indications for Naloxone, making it a more versatile tool in addiction management and public health.

Naltrexone has also been the subject of clinical trials that examine its role in managing conditions beyond addiction. Some studies are investigating its potential use in the treatment of **binge eating disorder**, **obesity**, and **depression**. By reducing cravings and reinforcing healthy behavior, Naltrexone may have broader applications in mental health and metabolic disorders. Additionally, new formulations of Naltrexone, such as **extended-release** versions, are being tested to improve patient adherence and outcomes.

New Indications for Naloxone and Naltrexone in Modern Medicine

Both Naloxone and Naltrexone are seeing their indications expanded to include conditions outside of addiction. This reflects the growing understanding of the brain's reward systems and the role of endorphins in a range of physiological and psychological processes.

1. **Naloxone in Non-Overdose Situations:** While Naloxone is widely known for its use in opioid overdose management, it is also being explored for use in **opioid-induced hyperalgesia** (increased pain sensitivity due to opioid use) and **opioid-induced constipation**. Naloxone may be used to mitigate these side effects by selectively blocking certain opioid receptors involved in these processes, offering a potential solution to common opioid-related side effects without compromising pain relief.

   Furthermore, researchers are investigating the use of Naloxone in treating opioid dependence, particularly as part of **opioid substitution therapy** (OST). OST traditionally uses medications like methadone or buprenorphine to prevent withdrawal symptoms and cravings. Naloxone could be combined with other therapies to address dependence and reduce the risk of misuse by blocking the euphoric effects of any illicit opioids taken.

2. **Naltrexone in Mental Health Treatment:** Naltrexone's ability to inhibit the effects of endorphins has led to exploration in various **psychiatric disorders**. Recent studies suggest that Naltrexone may be beneficial in treating conditions like **bipolar disorder, schizophrenia**, and **depression**. The theory behind its use in these conditions is that by inhibiting the brain's reward system, Naltrexone may help regulate mood fluctuations and reduce manic or depressive episodes. Additionally, Naltrexone's potential role in **obsessive-compulsive disorder (OCD)** and **addictive behaviors** related to non-substance use (such as gambling or internet addiction) is being actively researched. If successful, this could open up new avenues for treating a variety of mental health conditions where compulsive or addictive behaviors are central to the disorder.

3. **Expanding Use in Pain Management:** Naltrexone is being studied for its **analgesic properties** in managing chronic pain. Unlike opioid analgesics, Naltrexone works by blocking the opioid receptors without causing dependence or addiction. Low-dose Naltrexone (LDN) has been explored for **fibromyalgia, chronic fatigue syndrome, neuropathic pain**, and **inflammatory conditions**. The ability of Naltrexone to modulate the immune system and reduce inflammation is an area of active research, and early findings show promising potential for its use as an alternative to traditional pain management options.

Emerging Treatment Strategies Involving Opioid Antagonists

As the global opioid crisis continues to evolve, new treatment strategies involving opioid antagonists like Naloxone and Naltrexone are emerging to combat addiction and reduce overdose mortality rates. The concept of **early intervention**—administering Naloxone before an overdose occurs—has gained traction, especially for people at high risk of overdose, such as those with a history of opioid use or individuals recently discharged from detox programs.

In addiction recovery, **personalized medicine** is becoming a focal point, where treatments are tailored to the individual's genetic makeup, environmental factors, and addiction history. For example, genetic testing could help determine which patients are more likely to benefit from Naltrexone or Naloxone-based therapies, optimizing recovery plans and improving outcomes.

Moreover, the integration of **opioid antagonists** with **digital therapeutics** such as mobile apps and telemedicine is paving the way for a more accessible and integrated approach to addiction recovery. These technologies can monitor adherence, offer counseling, and provide support to individuals while they undergo treatment.

The Role of Naloxone and Naltrexone in the Global Health Landscape

Both Naloxone and Naltrexone have become essential tools in the global effort to address opioid addiction and overdose deaths. As countries struggle with the opioid epidemic, these medications are integral to public health strategies.

1. **Global Access to Naloxone:** Efforts to expand access to Naloxone have been a key focus for global health organizations, particularly in areas where opioid overdose deaths are most prevalent. Countries like the United States, Canada, and the United Kingdom have expanded Naloxone distribution programs, enabling first responders, healthcare workers, and even members of the general public to administer the life-saving drug. Global initiatives aimed at **harm reduction** have increased access to Naloxone, contributing to a reduction in overdose deaths.

2. **Addressing the Stigma of Addiction:** One of the challenges in expanding the use of Naloxone and Naltrexone is addressing the **stigma surrounding addiction**. People with substance use disorders often face societal discrimination, which can discourage them from seeking help. Public health campaigns and education are critical in reducing the stigma associated with addiction, particularly in making Naloxone and Naltrexone more widely accepted as part of recovery efforts. This requires collaboration between governments, healthcare professionals, and community organizations to create supportive, stigma-free environments for individuals in recovery.

3. **International Treatment Approaches:** Different countries have varying approaches to opioid addiction treatment, with some focusing more on **opioid substitution therapy** (e.g., methadone) and others, like Switzerland, integrating **Naloxone and Naltrexone** into broader harm reduction strategies. Understanding how different systems integrate opioid antagonists into their healthcare frameworks is important for optimizing recovery outcomes and reducing opioid-related harm on a global scale.

Conclusion: The Promise of Naloxone and Naltrexone in the Future of Addiction Medicine

Naloxone and Naltrexone have already revolutionized the way we treat opioid addiction and manage overdose. As ongoing research explores new applications for these medications—ranging from mental health treatment to chronic pain management—their role in healthcare will continue to expand. The future of opioid antagonist therapies lies in their integration with personalized medicine, behavioral therapy, and digital health tools, offering patients comprehensive and individualized treatment plans.

By expanding access to Naloxone and Naltrexone globally, addressing addiction stigma, and investing in research, we can improve health outcomes and save lives, offering new hope to individuals struggling with addiction and the broader global community. The continued evolution of these therapies represents a promising frontier in the battle against the opioid epidemic and other substance use disorders.

Chapter 12: Integrating Naloxone and Naltrexone into Clinical Practice

How to Incorporate Naloxone and Naltrexone into Treatment Regimens

As the use of Naloxone and Naltrexone expands across different areas of medicine, it is important for healthcare providers to understand how to effectively integrate these medications into treatment regimens. Whether in addiction recovery, pain management, or mental health treatments, proper incorporation of these drugs requires careful consideration of the patient's medical history, current health status, and specific therapeutic needs.

Naloxone in Emergency and Acute Care Settings:

Key Considerations:

- **Dosage:** Naloxone should be given in an appropriate dose based on the severity of the overdose. The initial dose is typically 0.4 to 2 mg, with repeated doses given every 2 to 3 minutes if necessary until the patient responds.

- **Administration Route:** Naloxone is commonly administered intramuscularly, intravenously, or intranasally, depending on the clinical setting and available resources.

- **Monitoring:** After administration, the patient should be carefully monitored for a return of overdose symptoms and for possible withdrawal symptoms. Continuous monitoring is crucial until the individual is stabilized.

Naltrexone in Addiction Recovery:

Key Considerations:

- **Initial Detoxification:** Patients must be opioid-free before starting Naltrexone to prevent withdrawal symptoms. For opioid addiction, this usually means a detoxification period of 7 to 10 days to clear opioids from the system.

- **Dosing:** Naltrexone is typically administered orally once daily or as an extended-release injection (Vivitrol) once a month. Oral Naltrexone is effective but may be less convenient for some patients compared to the long-acting injection, which ensures consistent adherence.

- **Behavioral Support:** While Naltrexone helps manage the physiological aspects of addiction, it should be used in conjunction with behavioral therapies such as Cognitive Behavioral Therapy (CBT), contingency management, or 12-step programs. These approaches address the psychological and emotional triggers of addiction and help individuals build the skills needed for long-term recovery.

- **Psychiatric Evaluation:** Naltrexone should be prescribed with caution in patients with a history of **severe depression, suicidal ideation**, or **mental health disorders**. These conditions should be assessed and managed before starting treatment.

Naltrexone in Chronic Pain Management:

Key Considerations:

- **Low-Dose Naltrexone (LDN):** Low-dose Naltrexone has gained attention for its potential in managing conditions like **fibromyalgia, chronic fatigue syndrome, neuropathic pain**, and **inflammatory conditions**. Dosing typically involves lower than standard doses, often around 1 to 5 mg daily, to help modulate the immune system and reduce inflammation.

- **Alternatives to Opioid Analgesics:** In patients who cannot tolerate traditional pain medications due to addiction or a history of opioid use, Naltrexone can be used in combination with other pain management strategies, such as non-opioid pain relievers, physical therapy, or acupuncture.

- **Monitoring for Pain Relief:** It is important to monitor the effectiveness of Naltrexone in pain management and adjust the treatment plan if necessary, especially in the case of complex pain syndromes.

Naloxone and Naltrexone in Co-occurring Disorders:

co-occurring mental health disorders

anxiety

depression

post-traumatic stress disorder (PTSD)

mental health care

Key Considerations:

- **Integrated Care Approach:** Treating both addiction and mental health simultaneously has been shown to improve outcomes. Combining Naloxone (in emergency situations) and Naltrexone (for long-term addiction recovery) with **psychiatric medications** (e.g., antidepressants, anxiolytics) and **psychotherapy** can provide a holistic treatment approach.
- **Monitoring for Mental Health Symptoms:** Naltrexone can sometimes lead to emotional side effects such as anxiety or irritability. Regular monitoring of mental health symptoms and adjustments to the treatment regimen may be necessary to manage these effects.

Training Healthcare Providers on the Proper Use of Opioid Antagonists

Effective integration of Naloxone and Naltrexone into clinical practice requires that healthcare providers are properly trained. Given the complexities of opioid use disorder, addiction, and the management of pain, training programs for healthcare providers should focus on:

1. **Recognizing Opioid Overdose:** Healthcare providers need to be able to identify the signs of opioid overdose (such as unresponsiveness, slow or irregular breathing, and pinpoint pupils) and respond appropriately by administering Naloxone.

2. **Opioid Antagonist Education:** Providers should be knowledgeable about the differences between Naloxone and Naltrexone, their pharmacodynamics, appropriate uses, side effects, and how to monitor patients for adverse reactions.

3. **Behavioral Health Integration:** Addiction treatment is most effective when it includes **behavioral interventions**. Healthcare providers should be trained to offer or refer patients for therapies that complement the pharmacological treatment, such as CBT, motivational interviewing, and family therapy.

4. **Patient Education:** Patients should be educated on the role of Naloxone and Naltrexone in their treatment, including how these medications work, potential side effects, and why they are being prescribed. This education is essential for improving adherence and engagement in treatment.

Guidelines for Monitoring Patient Response to Treatment

After Naloxone or Naltrexone is administered, it is important to monitor the patient's response and adjust the treatment regimen as necessary. Guidelines for monitoring include:

For Naloxone:

- **After overdose reversal:** Close observation is needed to ensure the patient's vital signs remain stable and that no further respiratory depression or overdose symptoms return.
- **Re-evaluation:** If Naloxone is administered, follow-up treatment may be required to manage any underlying conditions and to assess for the risk of relapse into overdose.

For Naltrexone:

- **Adherence and Side Effects:** Regular follow-ups should monitor for side effects such as nausea, mood changes, and liver function. Long-term use may also require monitoring for changes in emotional well-being and the potential need for psychiatric support.
- **Assessing Effectiveness:** Providers should regularly assess the effectiveness of Naltrexone in preventing relapse and managing cravings. Adjustments to treatment, including additional support for mental health, may be necessary.

Conclusion: Optimizing Treatment with Naloxone and Naltrexone

The integration of Naloxone and Naltrexone into clinical practice requires a comprehensive, patient-centered approach that includes careful assessment, individualized treatment planning, and a focus on both the physiological and psychological aspects of addiction. By combining these opioid antagonists with behavioral therapies and other medical interventions, healthcare providers can offer a holistic, effective approach to addiction recovery, pain management, and even mental health disorders.

Proper training, monitoring, and patient education are essential to ensure these medications are used safely and effectively. As we continue to expand the use of Naloxone and Naltrexone, they will play an increasingly critical role in improving public health and reducing the impact of opioid addiction and overdose worldwide.

Chapter 13: Clinical Case Studies on Naloxone and Naltrexone

Case Study 1: Successful Use of Naloxone in Emergency Settings

Patient Background: A 32-year-old male was found unresponsive at home by a family member. He had a known history of opioid use disorder (OUD) and had been prescribed prescription opioids for chronic pain management. The family member called emergency services after noticing his shallow breathing and blue lips.

Clinical Intervention: Upon arrival, the emergency responders assessed the patient and identified signs of opioid overdose, including pinpoint pupils and respiratory depression. Naloxone was administered intranasally (4 mg), and within 2 minutes, the patient's breathing returned to normal, and he regained consciousness.

Outcome: The patient experienced withdrawal symptoms, including agitation, sweating, and nausea, shortly after the Naloxone administration, which is common due to the sudden reversal of opioid effects. The patient was transported to the hospital, where he was monitored for several hours. Given his history of opioid dependence, he was provided with additional resources for addiction treatment and recovery, including referral to a detoxification facility.

Key Takeaways:

- **Rapid response with Naloxone** can prevent fatalities in opioid overdose situations, providing a critical window for intervention.
- **Withdrawal symptoms** are common after Naloxone administration and should be anticipated in patients with opioid dependence.
- Follow-up care, including **addiction treatment**, is essential to prevent future overdoses and support long-term recovery.

Case Study 2: Naltrexone in Long-Term Opioid Addiction Recovery

Patient Background: A 45-year-old female with a history of opioid use disorder (OUD) presented at an addiction treatment center. She had been sober for six months following an opioid detoxification program but was struggling with persistent cravings and a fear of relapse. She had previously used heroin and prescription opioids for pain and had undergone both inpatient and outpatient treatment in the past.

Clinical Intervention: The treatment team decided to start the patient on **oral Naltrexone** (50 mg/day) as part of her relapse prevention plan. Given her commitment to staying sober and her history of failed treatments, she was also paired with **cognitive-behavioral therapy (CBT)** to address her underlying psychological triggers for opioid use. The patient was educated on the potential side effects of Naltrexone, such as nausea, headache, and liver function issues, and advised to report any unusual symptoms.

Outcome: Over the next six months, the patient attended regular follow-up appointments to monitor progress. She reported a reduction in cravings and a significant improvement in her ability to cope with stress without turning to opioids. While she initially experienced some mild nausea upon starting the medication, these side effects subsided after the first few weeks. The combination of **Naltrexone** and **CBT** helped the patient maintain sobriety and successfully reintegrate into her family and work life.

Key Takeaways:

- **Naltrexone** can be an effective tool in **relapse prevention**, especially when combined with **behavioral therapy**.
- **Side effects**, including mild gastrointestinal discomfort, are common in the initial stages of treatment but can be managed with education and close monitoring.
- Ongoing **patient engagement** through follow-up appointments and psychological support is critical for sustained recovery.

Case Study 3: Naltrexone for Alcohol Use Disorder

Patient Background: A 50-year-old male with **alcohol use disorder (AUD)** had been in and out of recovery programs for over 15 years. Despite multiple attempts at sobriety, he struggled with intense cravings for alcohol and frequent relapses. His medical history was complicated by hypertension, and he had recently been diagnosed with mild cirrhosis of the liver due to long-term alcohol abuse.

Clinical Intervention: The patient was started on **Naltrexone (50 mg/day)** as part of a comprehensive treatment plan for his AUD. Given his liver issues, the prescribing physician ensured that the patient's liver function was stable before initiating Naltrexone. Additionally, the patient was referred to a **support group** for alcohol recovery and offered weekly **motivational interviewing** sessions to help him develop coping strategies.

Outcome: After six weeks of Naltrexone therapy, the patient reported a decrease in his alcohol cravings and a greater ability to control his drinking when exposed to social situations where alcohol was present. The patient continued with the prescribed dosage and remained engaged in his therapy. Over the next 12 months, he maintained abstinence from alcohol and continued attending support group meetings.

Key Takeaways:

- **Naltrexone** can significantly reduce **alcohol cravings** and prevent relapse when combined with other treatment strategies like therapy and support groups.
- In patients with **liver disease**, liver function should be closely monitored, as Naltrexone can have hepatotoxic effects, particularly at higher doses.
- **Psychosocial support** and **motivation-based therapies** play an important role in achieving long-term recovery from alcohol use disorder.

Case Study 4: Naloxone as Part of a Harm Reduction Strategy

Patient Background: A 28-year-old male with a long history of heroin use was enrolled in a harm reduction program after an opioid overdose three months ago. He was not yet ready for long-term addiction treatment but wanted to ensure that he had a safety net in case of another overdose. The patient was living in a community where opioid overdoses were common and wanted to prevent fatal overdose in the event of relapse.

Clinical Intervention: The harm reduction team provided the patient with **Naloxone (4 mg intranasal)** to keep on hand in case of emergency. The patient was educated on the signs of overdose and the proper steps to take when administering Naloxone, including calling for emergency medical help. He was also instructed to return for regular follow-ups, where he could access additional **Naloxone kits** and receive information on available addiction recovery resources.

Outcome: Two weeks later, the patient experienced a **heroin overdose** while using in a public restroom. His friends administered the Naloxone kit that he had received through the harm reduction program, and he regained consciousness shortly afterward. After being evaluated by emergency services, he was referred to an inpatient addiction treatment program, where he began a detoxification process followed by behavioral therapy.

Key Takeaways:

- **Naloxone** is a crucial component of **harm reduction strategies**, especially for individuals who are at high risk of overdose but are not yet ready for or able to commit to full addiction treatment.

- **Education on Naloxone use** and overdose prevention should be provided to individuals in high-risk environments, along with continued access to harm reduction resources.

- **Referral to treatment** after an overdose provides an opportunity to intervene at a critical moment, offering individuals the support they need to begin the recovery process.

Conclusion: Learning from Real-World Applications

These case studies illustrate the diverse ways in which Naloxone and Naltrexone can be utilized in clinical practice. From emergency overdose reversal to long-term addiction management, these medications are powerful tools in the fight against opioid addiction and alcohol use disorder. They have proven to be lifesaving in acute situations, but their true potential lies in their ability to support individuals through the recovery process.

Each case demonstrates the importance of personalized care, close monitoring, and the integration of pharmacological treatments with behavioral support. While Naloxone can provide immediate, life-saving intervention, Naltrexone offers a valuable strategy for relapse prevention and long-term recovery. Both medications are essential in a comprehensive treatment approach for addiction, highlighting the need for continued research, public education, and access to these crucial resources.

Chapter 14: Understanding the Legal Landscape of Naloxone and Naltrexone

Legal Regulations Surrounding the Use of Naloxone and Naltrexone

As opioid addiction and overdose rates have reached alarming levels globally, the role of medications like Naloxone and Naltrexone has become increasingly critical in public health initiatives. Both Naloxone and Naltrexone are classified as prescription medications, with each having distinct legal regulations and frameworks for use, prescribing, and distribution.

Naloxone Regulations:

life-saving medication

federal

state-level regulations

prescription-only medication

- **In the United States:** Naloxone is a **prescription medication** but can be prescribed to anyone at risk of opioid overdose, including family members, friends, and individuals with a history of opioid use disorder. **Standing orders** have been implemented in many states, allowing pharmacies to dispense Naloxone without the need for a patient-specific prescription. This policy is particularly important for **harm reduction** programs and emergency responders who are in direct contact with individuals at high risk of overdose.

- **Over-the-counter Access:** Some states and regions have passed laws allowing **over-the-counter Naloxone** distribution, particularly in places with high overdose rates. In 2019, for example, **New York State** made it easier to obtain Naloxone by allowing individuals to purchase it from participating pharmacies without a prescription.

- **Good Samaritan Laws:** In many U.S. states, **Good Samaritan Laws** provide legal protection to individuals who administer Naloxone during an overdose situation. These laws are designed to encourage bystanders to intervene without fear of prosecution, thus potentially saving lives.

Naltrexone Regulations:

opioid

alcohol

prescription medication

- **In the United States:** Naltrexone is a **prescription-only** medication for use in addiction treatment. Providers must ensure that patients are opioid-free before starting Naltrexone to avoid precipitating withdrawal symptoms. The **extended-release form** of Naltrexone (Vivitrol) is administered monthly and is closely monitored due to its potential side effects, such as liver toxicity.

- **Insurance Coverage and Access:** The use of Naltrexone in addiction recovery is generally covered by **health insurance** plans, but accessibility can vary widely. **State Medicaid programs** often include Naltrexone for addiction treatment, but out-of-pocket costs for extended-release formulations can be a barrier for some patients.

- **International Use:** In several countries, Naltrexone is prescribed not only for opioid addiction but also for alcohol use disorder (AUD). In the **European Union** and **Canada**, Naltrexone is available as a first-line treatment for individuals with AUD, and its use has been widely accepted as part of comprehensive addiction management strategies.

Prescription Guidelines and Access Issues

Despite the clear benefits of Naloxone and Naltrexone, there are **access barriers** that limit their full potential in combating opioid addiction and overdose deaths.

Naloxone Access Issues:

- **Economic Barriers:** Although the cost of Naloxone has decreased over the years, it remains relatively expensive in some regions, particularly the **extended-release intramuscular injection** forms. The high price point limits its accessibility in areas that need it most.

- **Healthcare Provider Education:** While many emergency providers and addiction specialists are well-trained in Naloxone use, general healthcare providers may be less familiar with the protocols for prescribing Naloxone. Increased **education and training** for healthcare providers can ensure that Naloxone is being prescribed to at-risk populations.

- **Pharmacy Access:** While Naloxone is available through some pharmacies without a prescription in certain states, it is still not universally accessible in all regions. Expanding **pharmacy access** and integrating Naloxone into **community harm reduction programs** are essential steps in increasing its availability.

Naltrexone Access Issues:

- **Eligibility for Naltrexone Treatment:** Patients seeking Naltrexone for opioid or alcohol addiction must first undergo detoxification to ensure they are opioid-free. This creates a barrier for individuals who are unable or unwilling to undergo detoxification, especially in emergency situations. Moreover, long-term access to Naltrexone, particularly the **extended-release formulation**, requires consistent healthcare visits and monitoring, which may not be feasible for all patients.

- **Insurance Coverage:** While insurance coverage for Naltrexone is generally good in the United States, certain **Medicare** and **Medicaid** plans may have restrictions or high copays for extended-release formulations, limiting access for low-income patients.

- **Global Disparities in Access:** In countries with less-developed healthcare infrastructures, access to Naltrexone remains limited. Governments and healthcare organizations must advocate for **international policy changes** to increase the availability of both Naltrexone and Naloxone in addiction treatment programs, particularly in countries with a high burden of opioid addiction.

Policies for Expanding Access to Naloxone in Emergency Situations

Given the growing prevalence of opioid overdoses, policymakers have focused on improving access to Naloxone as part of emergency **harm reduction strategies**.

1. **Standing Orders for Naloxone Distribution:** Standing orders have been established in many U.S. states, enabling pharmacies to distribute Naloxone to individuals at risk of opioid overdose without a prescription. This initiative has significantly improved the availability of Naloxone and allowed family members, friends, and caregivers to play an active role in overdose prevention.

2. **Public Health Campaigns:** Public health campaigns aimed at increasing awareness of Naloxone's life-saving potential have been crucial in reducing the stigma associated with opioid addiction. These campaigns promote **Naloxone accessibility** and train the public on how to recognize and respond to an overdose. In the U.S., the **CDC** and other health agencies have worked with local governments and non-profits to increase public awareness and distribute Naloxone kits. These programs also provide education on how to administer Naloxone in an emergency situation, which has proven effective in saving lives.

3. **Naloxone in Schools and Workplaces:** Policies are evolving to make Naloxone available in high-risk settings, such as schools and workplaces, where opioid overdoses may occur. As part of these initiatives, school nurses, teachers, and security personnel are often trained to recognize the signs of opioid overdose and administer Naloxone in an emergency.

Conclusion: Advocating for Expanded Access

The legal landscape surrounding Naloxone and Naltrexone has evolved significantly in recent years, driven by the opioid epidemic and the need for better solutions for addiction treatment and overdose prevention. While significant strides have been made to improve access, there is still much work to be done.

Efforts to expand **access to Naloxone** in emergency situations, improve **insurance coverage** for Naltrexone, and reduce **legal barriers** for individuals at risk of addiction are vital steps in addressing the public health crisis caused by opioid addiction. Increased **education for healthcare providers** and the public, as well as continued advocacy for **harm reduction policies**, will be crucial for realizing the full potential of these life-saving medications.

Ultimately, the future of Naloxone and Naltrexone hinges on **policy innovations**, continued public education, and the commitment of healthcare providers to improve patient access to these essential treatments. By ensuring widespread availability and support for these medications, we can continue to make significant strides toward overcoming the opioid crisis and improving the lives of those affected by addiction.

Chapter 15: Conclusion: Mastering Endorphin Inhibition in Medicine and Life

The opioid epidemic continues to pose one of the most pressing public health challenges worldwide. From overdose prevention to long-term addiction recovery, Naloxone and Naltrexone have emerged as invaluable tools in the arsenal against this crisis. These two medications, though distinct in their mechanisms of action, both work to inhibit endorphin activity, providing profound therapeutic benefits in the treatment of opioid addiction, overdose, and a range of other conditions.

This book has explored the **scientific principles** behind Naloxone and Naltrexone, their clinical applications, and the profound impact they have on both the **individual** and **society** at large. From emergency interventions to long-term recovery strategies, the integration of these opioid antagonists into clinical practice has revolutionized the treatment of opioid addiction and overdose, offering patients a path to recovery, a second chance at life, and in many cases, a **new sense of hope**.

Summary of Key Insights

1. **Endorphins and Their Role in the Body:** Endorphins, often referred to as the body's natural painkillers, play a critical role in regulating mood, pain, and emotional well-being. Understanding the physiological and psychological benefits of endorphins is essential for comprehending how **opioid antagonists** like Naloxone and Naltrexone can alter the body's reward system and influence addiction cycles.

2. **Mechanisms of Action:** Both Naloxone and Naltrexone work by **blocking opioid receptors**, specifically the **mu** receptor, which is responsible for the pleasurable effects of opioids. However, while **Naloxone** serves as an **emergency reversal agent** in overdose situations, **Naltrexone** is used as a **long-term maintenance therapy** in addiction recovery. Their ability to inhibit the rewarding effects of opioids and alcohol provides patients with the opportunity to break free from addiction and regain control over their lives.

3. **Therapeutic Applications and Clinical Benefits:** Naloxone's use in emergency overdose management has undoubtedly saved countless lives, offering a vital **lifeline** in the critical moments of opioid toxicity. In contrast, Naltrexone's role in relapse prevention and its ability to **reduce cravings** is foundational to long-term addiction recovery. By helping individuals remain abstinent from opioid use, Naltrexone has allowed many to sustain recovery and improve their quality of life.

4. **Pain Management and Mental Health:** The role of Naloxone and Naltrexone in **pain management** is particularly significant for individuals with a history of opioid addiction. By offering alternative approaches to managing chronic pain, these medications allow patients to avoid the risk of relapse while controlling their symptoms. Similarly, the use of Naltrexone in the treatment of psychiatric disorders such as depression and anxiety opens new avenues for addressing **mental health challenges** in conjunction with addiction treatment.

5. **Public Health Impact:** The accessibility of Naloxone has had a profound impact on public health, particularly in areas with high rates of opioid overdose deaths. **Harm reduction programs** and the wide distribution of Naloxone are essential strategies for saving lives and preventing overdose fatalities. These programs have demonstrated that public health interventions—when done with the **right tools** and **support systems**—can effectively reduce mortality rates and facilitate recovery.

6. **Legal and Ethical Considerations:** As Naloxone and Naltrexone continue to be utilized in emergency and addiction recovery settings, the **legal landscape** surrounding their distribution and use is evolving. Ensuring that Naloxone is easily accessible and that Naltrexone remains covered by insurance are critical steps in removing barriers to treatment. **Ethical concerns** around autonomy, consent, and the appropriate use of opioid antagonists also need ongoing discussion, particularly as these medications continue to be integrated into broader public health strategies.

7. **Global Perspective and Policy Advocacy:** Around the world, access to Naloxone and Naltrexone varies significantly. **Global health organizations** must continue to advocate for policies that prioritize the distribution of these medications, particularly in low-resource settings. The international effort to combat the opioid epidemic requires not only better access to these drugs but also a coordinated, multifaceted approach that includes **education**, **treatment**, and **prevention**.

The Future of Naloxone and Naltrexone

As we look to the future, the role of Naloxone and Naltrexone in the treatment of opioid addiction and overdose will likely continue to expand. **Research and clinical trials** are already underway to explore new indications for these medications, including their potential application in **chronic pain management, obesity,** and **psychiatric disorders**. Additionally, advancements in drug delivery systems—such as longer-acting injectable forms of Naltrexone—could improve adherence and make these medications more accessible to a broader population.

Beyond addiction recovery, the development of **opioid antagonist therapies** may offer new ways to address **reward-related** disorders, **mood disorders**, and even **neurodegenerative diseases**. The growing understanding of the **brain's reward pathways** and the role of endorphins in regulating emotions, pain, and addiction provides an exciting platform for expanding the clinical utility of these medications.

The Role of Healthcare Providers in Shaping the Future

Healthcare providers are on the front lines of the battle against opioid addiction and overdose, and their role in integrating Naloxone and Naltrexone into treatment regimens is crucial. Training providers in the effective use of these medications, as well as fostering a **non-judgmental approach** to addiction, will help improve patient outcomes and reduce the stigma associated with opioid use disorder.

Collaboration between healthcare professionals, addiction specialists, and public health officials will be essential in addressing the broader **social determinants of health** that contribute to addiction and overdose. By continuing to **advocate for better access** to opioid antagonists and **incorporating them into a comprehensive care model**, healthcare providers can help create an environment in which individuals struggling with addiction are empowered to seek treatment and live healthier, more fulfilling lives.

Conclusion: Empowering Lives through Endorphin Inhibition

Naloxone and Naltrexone are two of the most important medications in modern medicine for tackling the opioid crisis and beyond. Their ability to modulate the **reward system** in the brain and provide life-saving interventions marks a significant achievement in public health. By **mastering the principles of endorphin inhibition**, healthcare providers, policymakers, and communities can continue to transform the landscape of addiction treatment, pain management, and mental health.

As we move forward, the ongoing research, education, and advocacy surrounding these medications will play a pivotal role in addressing not just the opioid epidemic but also the broader challenges posed by addiction and mental health. Understanding the science, clinical applications, and social impact of Naloxone and Naltrexone will empower individuals and communities to not only **survive** but also **thrive** in the face of addiction and its complex challenges.

The journey to mastery in endorphin inhibition is not just about the science of medications but also about the **human stories of resilience, recovery, and transformation**. As we continue to advance our understanding and approach, we move closer to a world where addiction is seen not as a moral failure but as a treatable medical condition—and where every life saved brings us closer to a healthier, more compassionate society.

Chapter 16: Integrating Naloxone and Naltrexone into Clinical Practice

Introduction

The integration of Naloxone and Naltrexone into clinical practice is essential for optimizing treatment outcomes in patients dealing with opioid use disorder (OUD), opioid overdose, alcohol use disorder (AUD), and chronic pain management. Both of these medications—though functioning through different mechanisms—are key to addressing the complex nature of addiction and pain management, which are intertwined with the body's endorphin system. For healthcare professionals, understanding how to use these drugs effectively, manage their administration, and monitor patient outcomes is crucial to maximizing their therapeutic potential.

This chapter outlines strategies for incorporating Naloxone and Naltrexone into **clinical regimens**, presents **training recommendations** for healthcare providers, and provides **guidelines for monitoring patient responses**. By following these protocols, healthcare professionals can ensure that they are providing **comprehensive, evidence-based care** to individuals with addiction and pain management needs, thus promoting **better health outcomes**.

How to Incorporate Naloxone and Naltrexone into Treatment Regimens

Naloxone Use in Emergency Settings

emergency medicine protocols

harm reduction programs

- **Assess Overdose Risk:** Prior to prescribing or distributing Naloxone, it is important to identify individuals at high risk of opioid overdose, including those with a history of opioid use disorder, prescription opioid misuse, and other risk factors such as living in an area with high overdose rates.

- **Patient and Family Education:** Educate patients at risk for overdose—and their family members or caregivers—on the signs of an overdose and the steps for **Naloxone administration**. Training should include the use of **intranasal or intramuscular Naloxone** kits, including **how to administer the medication** and **when to seek emergency medical help**.

- **Prescribing Naloxone:** For those with a known history of opioid use disorder, **Naloxone can be prescribed** alongside opioid medications as part of a harm reduction approach. This is particularly important for patients transitioning from inpatient to outpatient care or those who are trying to manage **chronic pain** while reducing the risk of opioid misuse.

Naltrexone Use in Addiction Recovery

preventing relapse

cravings

comprehensive treatment plan

pharmacotherapy

behavioral therapy

- **Opioid Addiction Treatment:** Naltrexone should be prescribed after **detoxification** in individuals who are opioid-free for a minimum of 7–10 days. This is crucial to avoid precipitating withdrawal symptoms. The **oral form** (50 mg) is typically prescribed daily, while the **extended-release injectable form** (Vivitrol) can be administered monthly.

- **Alcohol Addiction Treatment:** Naltrexone is also highly effective for individuals with alcohol use disorder. It works by **reducing alcohol cravings** and the rewarding effects of drinking. The oral and injectable forms can be used, with some patients benefiting from the extended-release version to improve adherence.

- **Combination with Behavioral Interventions:** Combining Naltrexone with **cognitive behavioral therapy (CBT)** or **contingency management** has been shown to enhance treatment outcomes. **Support groups**, such as **12-step programs**, can also complement Naltrexone therapy by providing peer support during recovery.

Incorporating Naltrexone in Pain Management

chronic pain

opioid-blocking

- **Low-Dose Naltrexone (LDN):** There is growing interest in using **low-dose Naltrexone (LDN)** for conditions like **fibromyalgia, chronic fatigue syndrome**, and **multiple sclerosis**. Research has suggested that LDN works by **modulating the immune system** and reducing inflammation, providing a non-opioid option for pain management.
- **Guideline for Use in Pain:** For patients with a history of addiction or those at risk of developing opioid dependence, prescribing Naltrexone (even in low doses) can be an effective strategy for managing pain without triggering the **reward pathways** that lead to addiction.

Monitoring and Adjusting Treatment

optimal treatment outcomes

- **Naloxone Administration Monitoring:** When Naloxone is used in emergency settings, it is critical to **monitor patients closely after administration**, as they may experience **withdrawal symptoms**, including agitation, nausea, and vomiting. Although Naloxone has a short half-life, it is possible that **further doses** may be needed, especially if the opioids in the system are long-acting (e.g., fentanyl).

- **Naltrexone Response Monitoring:** Naltrexone's effectiveness in addiction recovery is often **measured by a decrease in cravings** and **relapse prevention**. Regular follow-up appointments should include **behavioral assessments** and check-ins on **mental health**, as some patients may experience side effects like **depression** or **anxiety**. Liver function tests are also important when prescribing Naltrexone, especially with the extended-release formulation.

Integrating Naloxone and Naltrexone in Combined Treatment Plans:

Naloxone

Naltrexone

comprehensive treatment plan

acute and long-term needs

short-term solution

sustained therapeutic intervention

pharmacological treatment

psychosocial interventions

Training Healthcare Providers on the Proper Use of Opioid Antagonists

Training on Naloxone Use:

proper use of Naloxone

- **Identifying signs of opioid overdose** (e.g., pinpoint pupils, shallow breathing, unresponsiveness).
- **Administering Naloxone** either intranasally or intramuscularly.
- **Post-administration care**, such as ensuring the patient is in a **recovery position**, monitoring **vital signs**, and ensuring they are taken to a healthcare facility.

Training on Naltrexone Use:

- **Patient selection criteria**, ensuring that patients are opioid-free before starting Naltrexone therapy.
- **Managing withdrawal symptoms** that may arise when starting Naltrexone.
- **Educating patients** on the importance of adherence and integrating Naltrexone with **behavioral therapies** for best results.

Guidelines for Monitoring Patient Response to Treatment

1. **Monitoring Naloxone Use:** After administering Naloxone, patients should be carefully monitored for at least 30 minutes to 1 hour, as opioid effects may return after Naloxone's short duration of action. The patient's **vital signs**, especially **breathing and consciousness**, should be closely observed.

2. **Monitoring Naltrexone Use:** Regular check-ups are essential to monitor the effectiveness of Naltrexone in preventing relapse and reducing cravings. **Psychosocial assessments** should also be part of the treatment plan, as mental health factors may influence the patient's adherence to Naltrexone.

Conclusion

Incorporating **Naloxone** and **Naltrexone** into clinical practice involves understanding their pharmacological properties, the best ways to administer them, and how to monitor patients for both effectiveness and potential side effects. By providing comprehensive care that includes medication, patient education, and psychosocial support, healthcare providers can empower patients to take control of their recovery and reduce the burden of opioid addiction and overdose deaths.

Ultimately, integrating these opioid antagonists into clinical regimens is not only about treating addiction but also **transforming lives**. With proper training, monitoring, and a patient-centered approach, **Naloxone** and **Naltrexone** can be used to **create lasting change** for individuals struggling with addiction, improving their health, well-being, and prospects for a future free from opioid dependence.

Chapter 17: Understanding the Legal Landscape of Naloxone and Naltrexone

Introduction

As the opioid epidemic continues to claim lives across the world, the importance of medications like Naloxone and Naltrexone cannot be overstated. These drugs, both of which work by inhibiting the effects of endorphins through opioid receptor antagonism, have proven critical in saving lives, reducing overdose deaths, and aiding in the recovery from opioid addiction. However, despite their effectiveness, the use and distribution of Naloxone and Naltrexone are subject to a complex web of **legal regulations** and **policies** that can significantly impact patient access to these life-saving treatments.

This chapter examines the legal framework surrounding the use of Naloxone and Naltrexone, the prescription guidelines, and the regulatory hurdles that healthcare providers, patients, and policy-makers must navigate. In particular, we will explore the **challenges** and **opportunities** that come with expanding access to these medications, including **public health policies**, **laws related to opioid prescribing**, and **efforts to combat stigma**.

Legal Regulations Surrounding Naloxone and Naltrexone

Naloxone Access and Distribution Laws

public health initiatives

- **Over-the-Counter Naloxone:** In some places, Naloxone is available over-the-counter (OTC) without the need for a prescription, especially in harm reduction programs. In **many U.S. states**, pharmacists are legally authorized to dispense Naloxone to individuals at risk of opioid overdose or to their family members, even if they do not have a prescription from a physician. This approach has been endorsed by many **public health agencies** to expand access and reduce overdose deaths.

- **Standing Orders for Naloxone Distribution:** A **standing order** allows **physicians** or **public health officials** to prescribe Naloxone to an entire population, such as people at risk of opioid overdose, without the need for individual prescriptions. This practice has gained significant traction as part of **harm reduction strategies**. Many states in the U.S. have passed laws that allow such standing orders, broadening access to Naloxone for individuals, community-based organizations, and pharmacies.

- **Good Samaritan Laws:** Several U.S. states have implemented **Good Samaritan Laws** that protect individuals from criminal liability when they administer Naloxone in an emergency overdose situation. These laws have helped encourage people to intervene in overdose emergencies without fear of prosecution, thereby increasing the chances of saving lives.

2. While Naloxone's accessibility has increased in many places, some **countries and states** still face regulatory barriers that prevent widespread distribution. These barriers often stem from **insurance coverage issues**, **cost concerns**, and **lack of knowledge** about the drug's importance. Public health advocacy is necessary to remove these barriers and ensure **widespread access** to Naloxone.

3. **Naltrexone Access and Prescribing Guidelines** Naltrexone is a **prescription medication**, and its distribution is subject to more traditional prescription regulations. However, like Naloxone, Naltrexone plays a significant role in the **treatment of opioid and alcohol use disorders**, making it essential for healthcare providers to understand the legal requirements and the conditions under which it can be prescribed.

- **Prescribing Naltrexone for Addiction Treatment:** Naltrexone is most commonly prescribed for individuals with opioid and alcohol use disorders, where it functions as a relapse prevention agent. Providers must ensure that patients are **opioid-free** before initiating Naltrexone therapy. In many countries, this medication is only available through **licensed prescribers** (such as physicians or addiction specialists), and providers must carefully screen patients for suitability. For opioid-dependent patients, **detoxification** is required before beginning treatment to avoid precipitating withdrawal symptoms.

- **Extended-Release Naltrexone (Vivitrol):** The injectable form of Naltrexone, **Vivitrol**, is a long-acting formulation that is typically administered every 4 weeks. It has the advantage of improving **adherence** to treatment, but it also requires healthcare providers to manage the **administration** and monitor patients for side effects. Regulations concerning the administration of injectable Naltrexone vary depending on the healthcare system and whether patients are insured.

- **Off-Label Use of Naltrexone:** While Naltrexone is approved for opioid and alcohol use disorders, it is sometimes used off-label to treat other conditions, including **binge eating disorder, obesity,** and **chronic pain**. Prescribing Naltrexone for these conditions requires a **comprehensive understanding** of the potential risks and benefits, as well as legal and ethical considerations regarding off-label prescribing

Insurameetcoverage and Cost Considerations

insurance coverage

complex reimbursement systems

- **Naloxone and Insurance Coverage:** Many **public health programs**, including Medicaid and Medicare in the United States, offer **coverage for Naloxone** as part of emergency care or harm reduction programs. However, access to Naloxone through private insurance can vary, and some individuals may find themselves **unable to afford** the medication, especially if they do not qualify for government assistance programs. Advocating for better insurance coverage and making Naloxone **widely available** in pharmacies are key steps toward improving public health outcomes.

- **Naltrexone and Cost Concerns:** Naltrexone, particularly the extended-release injectable form, can be **cost-prohibitive** for many patients, especially those without insurance. Ensuring access to affordable Naltrexone is crucial, as it plays a significant role in **preventing relapse** in addiction treatment. Programs that provide **subsidies** or financial assistance for these medications are critical in ensuring equitable access to treatment.

Policies for Expanding Access to Naloxone in Emergency Situations

As Naloxone continues to be one of the most effective tools in preventing opioid overdose deaths, **policy changes** aimed at expanding access are essential. Government action at both **local** and **national levels** can create systemic changes that help make Naloxone widely available.

Policy Recommendations:

wider distribution

accessibility

- **Distribution through harm reduction programs** (needle exchange programs, community pharmacies, and public health clinics).
- **Making Naloxone widely available** in **schools**, public places, and at events where opioid use is a concern.
- **Allowing paramedics** and first responders to carry and administer Naloxone without medical oversight in certain emergency situations.
- **Government subsidies** to reduce the cost of Naloxone for uninsured individuals.

Stigma and Its Impact on Access to Treatment

One of the most persistent barriers to accessing opioid-related treatments like Naloxone and Naltrexone is the **stigma** associated with addiction. People suffering from addiction often face **social discrimination**, which can make them hesitant to seek treatment or engage with healthcare providers.

- **Reducing Stigma:** Public health campaigns that promote **understanding** and **empathy** for those with substance use disorders can go a long way in **breaking down barriers**. By reframing addiction as a **chronic, treatable medical condition** rather than a moral failing, healthcare professionals can help alleviate the shame that often prevents individuals from seeking help.

- **Training Healthcare Providers:** Medical professionals must be trained to provide care in a **nonjudgmental manner** and offer support for individuals seeking treatment. Encouraging **open dialogue** about addiction and its treatment options is vital in ensuring that patients feel **comfortable and empowered** to access Naloxone and Naltrexone when they need it most.

Conclusion

The legal landscape surrounding the use of Naloxone and Naltrexone is complex, but it is steadily evolving to meet the demands of the opioid crisis and broader addiction treatment needs. By addressing legal barriers, expanding insurance coverage, and tackling the stigma associated with addiction, **policy-makers** and **healthcare professionals** can increase access to these life-saving medications. The future of addiction treatment depends on continuing to **advocate for greater access**, **better policies**, and **educational programs** that promote understanding, ensure affordability, and encourage compassionate care for all individuals affected by opioid use and addiction.

Chapter 18: The Ethical Considerations of Endorphin Inhibition

Introduction

The use of Naloxone and Naltrexone has revolutionized the way we approach the treatment of opioid use disorder (OUD), opioid overdose, and alcohol dependence. While these medications offer life-saving benefits, they also raise important ethical questions that deserve careful consideration. The ethical implications of endorphin inhibition—particularly in the context of addiction treatment, overdose reversal, and long-term recovery—are multi-faceted and require healthcare providers, policymakers, and society at large to balance public health, individual autonomy, and social responsibility.

This chapter explores the **ethical challenges** associated with the use of Naloxone and Naltrexone, as well as broader considerations regarding their roles in public health, criminal justice, and social policy. By examining these issues, we can better understand the broader ethical landscape surrounding endorphin inhibition and its implications for addiction medicine and beyond.

1. The Ethics of Naloxone Administration in Overdose Situations

Naloxone has become a critical tool in the fight against the opioid epidemic. Its ability to reverse opioid overdose and save lives in emergency situations has led to widespread adoption of take-home Naloxone kits, distributed by community organizations, harm reduction programs, and even pharmacies in some regions.

However, the widespread distribution and use of Naloxone bring up several **ethical questions** related to its **appropriateness** and **impact**:

- **Enabling Addiction or Saving Lives?**

  One of the key ethical dilemmas in Naloxone distribution is the question of whether it could be seen as enabling continued opioid misuse. Critics argue that making Naloxone easily accessible might reduce the urgency for individuals to seek long-term recovery solutions, potentially sustaining the cycle of addiction. On the other hand, proponents contend that Naloxone should be viewed as a **harm reduction strategy**—a tool to save lives, not as a substitute for recovery. This ongoing debate is a crucial point in determining how to integrate Naloxone effectively into public health strategies.

- **Access to Naloxone:**

  Ethical issues also arise around **access to Naloxone**. Should it be universally available to anyone at risk of opioid overdose, including those without a history of addiction? How do we ensure that people who need Naloxone most—often marginalized populations, including people in **high-risk communities** or **those in prison systems**—can access it? Ensuring equity in access to life-saving medications, like Naloxone, is a key ethical concern, and decisions about distribution and accessibility must be handled with sensitivity to social inequalities.

- **Public vs. Private Health Concerns:**

  Some argue that widespread Naloxone distribution is a **public health initiative**, necessary to address the opioid crisis, which has become a societal problem. Others question whether public funds should be used for overdose prevention, especially when some individuals may continue using opioids despite the risk. The tension between **individual responsibility** and **societal responsibility** for health care interventions is a crucial ethical debate.

2. Ethical Considerations in Using Naltrexone for Addiction Treatment

Naltrexone is used to help manage both opioid and alcohol dependence, blocking the euphoric effects of these substances and reducing cravings. While its use has been a **game-changer** in addiction treatment, several ethical considerations arise in its application:

- **Autonomy and Informed Consent:**

  For patients using Naltrexone in the treatment of opioid or alcohol addiction, ensuring **informed consent** is vital. Some patients may be prescribed Naltrexone as part of a comprehensive treatment plan that includes counseling, therapy, and lifestyle changes. However, there is a risk of **coercion** in some cases, particularly when the treatment is part of **court-mandated** or **involuntary recovery programs**. Patients must be made fully aware of the drug's potential side effects and its impact on their ability to use opioids or alcohol. The ethical principle of **patient autonomy**—allowing individuals to make informed decisions about their treatment—should always guide clinical practice.

- **Long-Term Use and Quality of Life:**

  Another ethical consideration is the **long-term use of Naltrexone**. Some patients may need to remain on Naltrexone for extended periods to prevent relapse, but the drug can have side effects, including gastrointestinal issues, insomnia, and fatigue. In such cases, healthcare providers must balance the benefits of preventing relapse with the potential for **reduced quality of life** due to these side effects. Regular monitoring and adjustment of treatment plans are necessary to ensure that the patient's well-being is maintained over time.

- **Ethical Dilemmas in Maintenance Treatment:**

  Naltrexone can be used as part of a **maintenance therapy** strategy to support long-term recovery from addiction. The ethical question arises: is it acceptable to use a drug to keep a patient in a chemically induced state of sobriety? While this approach can reduce cravings and promote abstinence, it can also raise concerns about **pharmacological dependence** and the loss of **personal agency**. Are we enabling a form of sobriety that doesn't empower the individual to make changes on their own?

3. Ethical Considerations in Addressing Addiction as a Disease vs. a Moral Failing

The growing use of Naloxone and Naltrexone has prompted society to confront deeper ethical questions about the nature of addiction itself. Is addiction a disease—one that requires medical intervention—or a moral failing that deserves punishment and shame?

- **Addiction as a Disease:**

  Most modern medical frameworks, including those embraced by the **American Medical Association (AMA)** and the **National Institute on Drug Abuse (NIDA)**, treat addiction as a **chronic, relapsing brain disease** that alters the way the brain functions. Understanding addiction as a disease rather than a moral failing is essential for developing compassionate treatment programs and effective interventions. Naltrexone and Naloxone both represent **medical interventions** that address the physiological and neurological aspects of addiction, rather than framing the person struggling with addiction as simply **weak-willed** or **morally deficient**.

- **The Role of Society in Addiction Treatment:**

  Ethical questions also arise about the role of society in supporting or stigmatizing individuals with addiction. **Social policies** that criminalize drug use or stigmatize people who use substances can hinder access to treatment and create a hostile environment for recovery. By framing addiction through the lens of **disease** and focusing on **harm reduction**, healthcare providers and policymakers can shift societal attitudes and create a more supportive environment for those seeking recovery.

Conclusion

The use of Naloxone and Naltrexone is a critical part of modern addiction medicine and overdose prevention. However, as their applications continue to expand, the ethical issues surrounding these drugs will need to be carefully examined and addressed. From ensuring **equitable access** to these life-saving treatments to navigating the complexities of **patient autonomy** and **informed consent**, the ethical landscape of endorphin inhibition is both challenging and essential to address. By considering these ethical principles, we can foster a more compassionate and effective approach to addiction treatment, one that truly serves the needs of patients while balancing societal concerns.

Chapter 19: The Future of Endorphin Inhibition: Advancements in Naloxone and Naltrexone Research

Introduction

The landscape of addiction treatment and pain management has been forever changed by medications like Naloxone and Naltrexone. Both are powerful tools that act through the inhibition of endorphins to help manage opioid use disorder (OUD), prevent overdose deaths, and assist in long-term recovery. However, despite their widespread use, the future of these drugs is far from static. As research into endorphin inhibition and opioid receptors continues to evolve, new advancements and potential applications are emerging, paving the way for more effective and targeted treatments.

This chapter explores the cutting-edge research that is shaping the future of **Naloxone** and **Naltrexone**, focusing on innovative therapeutic strategies, potential new indications for these medications, and the ongoing quest for better solutions to the opioid crisis.

Innovations in Naloxone: Expanding Its Use

Naloxone has proven to be a **life-saving intervention** in emergency overdose situations. Yet, as the opioid epidemic grows more complex, researchers are exploring ways to make Naloxone even more effective in a wider array of circumstances. Several avenues of innovation are currently being explored in clinical trials and research labs:

1. **Long-Acting Formulations**: Currently, Naloxone is typically administered through injectable or nasal spray forms, which are short-acting. Long-acting formulations of Naloxone are being developed to provide prolonged protection against overdose in at-risk individuals, potentially reducing the need for frequent re-dosing in emergency situations. This could be particularly valuable for individuals with high opioid tolerance or those who are transitioning from acute overdose treatment to long-term care.

2. **Naloxone Delivery Devices**: While the intranasal spray is an effective method for emergency use, researchers are investigating more user-friendly delivery systems, including **auto-injectors** and **implants** that provide continuous or longer-lasting effects. These innovations aim to make Naloxone more accessible, especially for individuals at high risk of overdose, and for use by non-medical personnel in emergency situations.

3. **Naloxone as Part of a Comprehensive Treatment Program**: Some research is exploring how Naloxone can be integrated more fully into long-term treatment strategies for opioid addiction. Rather than just being a tool for emergency reversal, Naloxone may become a part of a **harm-reduction approach**, used alongside behavioral therapies, medication-assisted treatment (MAT), and other recovery resources.

Advancements in Naltrexone: Broader Applications

While Naltrexone has primarily been used to treat opioid addiction and alcohol use disorder, its potential applications are expanding as well. Researchers are examining how Naltrexone could be used for other conditions, exploring its broader therapeutic role in addiction treatment and beyond.

1. **Naltrexone in Treating Other Substance Use Disorders**: Although Naltrexone is most commonly used for opioid and alcohol dependence, there is growing interest in its potential for treating other forms of addiction, such as **gambling**, **nicotine**, or even **food addiction**. Some studies have suggested that Naltrexone may be effective in reducing compulsive behaviors by modulating the brain's reward system, which is similar to the mechanism it uses to reduce alcohol and opioid cravings.

2. **Naltrexone and Pain Management**: Unlike traditional opioids, which activate the opioid receptors to relieve pain, Naltrexone works by blocking those same receptors, potentially offering an alternative route for pain management. Some researchers are investigating **low-dose Naltrexone (LDN)** for its ability to modulate the immune system and reduce pain and inflammation in conditions like **fibromyalgia**, **chronic pain**, and **autoimmune disorders**. Although LDN is not yet approved for these uses, early clinical trials have been promising, suggesting that Naltrexone could offer a non-addictive, non-opioid solution for managing chronic pain.

3. **Combination Therapies with Naltrexone**: Future research into **combination therapies** is exploring how Naltrexone can be used alongside other medications or treatments to increase its effectiveness. For example, pairing Naltrexone with **buprenorphine** (another medication used in opioid addiction treatment) could potentially enhance recovery outcomes by targeting different aspects of addiction. Moreover, using Naltrexone in conjunction with **cognitive-behavioral therapy** (CBT) and other therapeutic modalities may improve patient adherence and long-term success rates.

Personalized Medicine: Tailoring Treatments to the Individual

As the understanding of endorphin inhibition and opioid receptor activity deepens, the focus is shifting toward **personalized medicine**. In the future, patients may receive treatments that are specifically tailored to their individual genetic makeup, health history, and addiction profile. This could lead to more effective use of both Naloxone and Naltrexone, as physicians would be able to select the right medication, dose, and delivery method based on the patient's unique needs.

Genomic studies are already revealing that certain genetic variations may influence how individuals respond to opioid medications and addiction treatments. By analyzing a patient's genetic profile, doctors may soon be able to predict which medications will be most effective for them, minimizing side effects and improving treatment outcomes.

Global Access to Naloxone and Naltrexone

One of the most critical factors in addressing the opioid epidemic on a global scale is ensuring access to life-saving medications like Naloxone and Naltrexone. In many countries, especially in low-income regions, these medications remain limited or inaccessible due to cost, regulatory barriers, and lack of awareness.

Efforts are being made to increase the availability of Naloxone and Naltrexone globally. The **World Health Organization (WHO)** has recognized Naloxone as an essential medicine for opioid overdose reversal, and numerous non-governmental organizations (NGOs) are working to distribute Naloxone in countries hardest hit by the opioid crisis. Increasing global access to both Naloxone and Naltrexone will require continued advocacy, financial investment, and changes to local policies to ensure that these medications can be distributed and used effectively.

Conclusion: Looking Ahead

The future of Naloxone and Naltrexone is bright, with ongoing research continuing to unlock new therapeutic potentials. While challenges remain, particularly in terms of global access and the need for more personalized treatment approaches, these medications stand at the forefront of the battle against the opioid crisis and substance use disorders.

As scientific innovation continues, it is likely that new formulations, combination therapies, and treatment paradigms will emerge, offering hope to millions who suffer from addiction and pain. In the coming decades, we may look back at the progress made with these medications as a pivotal moment in the fight against addiction, one that saved countless lives and forever changed the landscape of medicine.

Chapter 20: Final Thoughts: Mastering the Balance of Endorphin Inhibition

Introduction

As we conclude this exploration into Naloxone, Naltrexone, and their profound impact on endorphin inhibition, it becomes evident that these two medications are not merely tools in the fight against opioid overdose and addiction. They represent pivotal advancements in our understanding of the body's opioid system and its role in both the relief and regulation of pain, as well as in the management of substance use disorders.

While **Naloxone** has proven to be a game-changer in reversing the lethal effects of opioid overdose, **Naltrexone** offers a longer-term solution to help individuals maintain recovery and rebuild their lives after addiction. Together, these medications embody the power of medical intervention to correct imbalances in the body's natural systems and restore homeostasis, providing both **immediate** and **sustained** benefits.

The Importance of Endorphin Regulation in Medicine

At the heart of this book lies the concept of **endorphin regulation**—understanding how endorphins interact with our opioid receptors and influence mood, pain perception, and emotional well-being. Endorphins are, in essence, the body's natural "painkillers," and their role in our lives extends far beyond addiction and overdose. By mastering endorphin inhibition with drugs like Naloxone and Naltrexone, we not only save lives in moments of crisis but also offer individuals a chance to regain control over their bodies and minds.

This regulation of endorphins is a delicate balance, one that requires **scientific precision**, **medical expertise**, and **human empathy**. These drugs allow us to manage an individual's physiological response to pain and stress, but they also provide a framework for treating addiction—an inherently complex and multifactorial condition that intertwines physical, psychological, and social dimensions.

Looking Toward the Future

The future of **Naloxone** and **Naltrexone** lies in continued research and clinical innovation. As we continue to deepen our understanding of the opioid system, new medications and treatment paradigms are likely to emerge. These will likely focus on:

- **Targeted therapies** to optimize endorphin inhibition in ways that minimize side effects.
- **Personalized medicine** approaches, where treatments are tailored to the individual's genetic makeup, response to drugs, and specific health needs.
- **Combination therapies**, where Naloxone or Naltrexone is used in conjunction with other interventions, such as cognitive-behavioral therapy (CBT) or mindfulness practices, to address the psychological aspects of addiction.

Further advancements may also focus on improving the **accessibility** of Naloxone, ensuring it is available to individuals who are at high risk of overdose, particularly in underserved communities. Widespread use of **take-home Naloxone kits** has the potential to save thousands of lives, especially when combined with education and harm reduction programs.

The Role of Society in Addiction Recovery

It is also important to recognize that the success of Naloxone and Naltrexone as tools for combating addiction and opioid overdose is not solely dependent on the medication itself. The broader **societal context**—which includes **mental health support, family involvement, community resources**, and **recovery programs**—plays an integral role in long-term recovery. While these medications are vital, they are most effective when accompanied by a holistic approach to addiction treatment.

We must continue to fight against the stigma surrounding addiction and foster a more compassionate, supportive environment for those struggling with substance use disorders. The medical community, policy-makers, and the general public all have roles to play in creating an environment where individuals can receive the care, compassion, and treatment they need to rebuild their lives.

Concluding Thoughts

Ultimately, mastering endorphin inhibition through Naloxone and Naltrexone represents not just a medical breakthrough, but a philosophical one. It underscores the importance of understanding and managing the delicate balance between **relief** and **addiction**, between **pain** and **pleasure**, and between **healing** and **dependence**. By carefully integrating these medications into clinical practice and society, we move closer to a world where opioid overdoses are preventable, addiction is treatable, and recovery is possible for all.

Endorphins are a powerful force in the body, but through the strategic use of Naloxone and Naltrexone, we gain the ability to regulate and modulate these forces in ways that benefit individual health, public safety, and overall well-being. Mastering endorphin inhibition isn't just about controlling a physiological response; it's about restoring balance, healing, and offering hope.

Thank you for joining me on this journey through the science, application, and potential of Naloxone and Naltrexone. The work does not end here—it continues with the ongoing efforts of healthcare professionals, researchers, policy-makers, and individuals striving to improve lives and reduce the burden of opioid addiction. The future is one of continued learning, compassion, and innovation.

Chapter 21: Expanding the Scope: Beyond Opioid Use and Addiction

Introduction

While Naloxone and Naltrexone are most commonly associated with opioid use disorder (OUD) and opioid overdose reversal, the potential applications of these medications extend far beyond the treatment of addiction. As researchers continue to explore the intricate mechanisms of the opioid system and endorphin regulation, new avenues for the use of these drugs are emerging, especially in areas of pain management, mental health, and even non-addiction-related behaviors.

This chapter will discuss the **expanding scope** of Naloxone and Naltrexone, examining how their therapeutic potential might be applied to **broader medical conditions**, **behavioral health**, and **innovative treatments** in the future. By exploring their use outside the traditional realm of addiction medicine, we can gain a deeper appreciation for these drugs and their lasting impact on both medical practice and public health.

1. Naloxone Beyond Addiction: Exploring New Therapeutic Uses

Although Naloxone is primarily used in overdose situations to rapidly reverse the effects of opioid toxicity, its potential applications are being expanded in several areas of medicine. Researchers have begun to explore its use in treating **chronic pain, post-surgical recovery**, and even **neurological conditions**.

- **Chronic Pain Management:** Chronic pain remains a major challenge in healthcare, often leading to opioid prescriptions that contribute to the very crisis Naloxone was designed to mitigate. Interestingly, some studies have suggested that **Naloxone** could play a role in **preventing opioid-induced hyperalgesia** (increased sensitivity to pain). By inhibiting certain opioid receptors, Naloxone might help to reduce pain sensitivity and **improve pain management** without the need for higher doses of opioids.

  Moreover, ongoing studies are exploring the use of Naloxone in combination with other pain relievers to enhance **analgesic effects** while reducing the potential for **dependence and overdose**. This could potentially offer a safer, long-term solution for managing chronic pain without resorting to opioids.

- **Neurodegenerative Diseases and Brain Injury:** Emerging research has also pointed to Naloxone's potential neuroprotective effects, especially in the context of **brain injuries** and **neurodegenerative diseases** such as **Parkinson's** and **Alzheimer's disease**. Naloxone's ability to modulate neuroinflammation and its role in regulating brain cell death make it an exciting candidate for future research into **neurodegenerative treatments**.

- **Psychiatric Applications:** Recent studies have begun to examine the potential of Naloxone in treating **depression**, **anxiety**, and other mood disorders. By blocking opioid receptors in the brain, Naloxone might provide insight into the relationship between opioid signaling and **emotional regulation**. Early research into this area is still in its infancy, but there is growing evidence that opioid antagonists could be valuable additions to the psychiatric toolbox.

2. Naltrexone in Non-Addiction Treatments: A Broader Medical Role

Naltrexone has long been used as a treatment for **alcohol dependence** and **opioid use disorder**, but researchers are now investigating its potential in treating a variety of non-addiction-related conditions.

- **Weight Management and Obesity:** One of the most promising new uses for Naltrexone is in the treatment of **obesity**. It has been shown to affect **appetite regulation** by altering the pathways in the brain that control hunger and satiety. By blocking opioid receptors, Naltrexone may help curb the **binge eating behaviors** that contribute to weight gain and obesity. When combined with other medications like **bupropion**, Naltrexone has been marketed as an effective option for patients struggling with **overweight** and **obesity**.

- **Autoimmune Diseases and Inflammation:** Some researchers are investigating Naltrexone's potential role in managing **autoimmune diseases** and **chronic inflammation**. **Low-dose naltrexone (LDN)** is an emerging treatment for conditions like **multiple sclerosis, fibromyalgia, Crohn's disease**, and **rheumatoid arthritis**. Its effects are believed to be tied to its ability to modulate the immune system and reduce excessive inflammation, although more studies are needed to confirm its long-term effectiveness in these areas.

- **Pain Management:** Similar to Naloxone, Naltrexone is also being explored as a potential **pain management** tool. By blocking opioid receptors, Naltrexone can prevent opioid-induced pain relief from escalating into dependence. Research into the use of Naltrexone for **chronic pain conditions**, particularly those that are not easily treated with traditional analgesics, is still ongoing but holds promise for alternative, safer treatments.

3. Potential Risks and Considerations in Expanding Use

While the expanding applications of Naloxone and Naltrexone are exciting, they do come with challenges and risks. As these medications are introduced into new therapeutic areas, it is essential that careful clinical trials are conducted to assess their **safety, efficacy**, and **long-term effects** in different patient populations.

For instance, Naltrexone's ability to block opioid receptors can lead to **withdrawal symptoms** in individuals who are physically dependent on opioids, potentially making its use problematic in patients who have not been adequately detoxified. Additionally, while low-dose naltrexone (LDN) has shown promise for several autoimmune diseases, its off-label use should be monitored to prevent misuse or improper self-medication.

Similarly, while Naloxone has been shown to be a **safe and effective** means of reversing opioid overdose, its use in non-overdose situations (such as pain management) must be approached cautiously, as it could have unforeseen effects on patients who are on other pain management regimens.

Conclusion

The growing body of research surrounding Naloxone and Naltrexone suggests that their role in medicine will continue to evolve, with expanding applications in areas far beyond addiction treatment. From **pain management** to **autoimmune disease therapy**, these medications hold significant promise for improving a range of conditions.

As we continue to explore and understand the broader impact of endorphin inhibition, Naloxone and Naltrexone will likely become key players in **future treatments**, offering new solutions to longstanding medical challenges. Their potential is vast, and with continued research, the possibilities for their use are only beginning to be realized.

Chapter 22: Personalized Medicine and the Future of Naloxone and Naltrexone

Introduction

As we continue to make strides in understanding the complexities of the brain, the body's opioid system, and how endorphins play a crucial role in various physiological processes, the future of **Naloxone** and **Naltrexone** is shifting toward more **personalized approaches** in medicine. Personalized medicine, sometimes referred to as precision medicine, tailors medical treatment to individual characteristics, including genetic makeup, lifestyle, and the unique biology of the patient. This approach is proving to be a game-changer in many areas of healthcare, and the use of Naloxone and Naltrexone is no exception.

In this chapter, we explore how **personalized medicine** could enhance the **effectiveness** of these medications, leading to **improved outcomes** in addiction treatment, pain management, and even in areas not yet widely studied. From **genetic factors** to **biomarker-guided treatment**, we will examine how these evolving fields may shape the future of endorphin inhibition therapies.

Genetic Variations in Opioid Receptors

A key component of personalized medicine is understanding that individuals have unique genetic profiles that affect how their bodies respond to various drugs. When it comes to opioids and endorphin regulation, genetic differences in opioid receptors (mu, kappa, and delta) can significantly influence a person's sensitivity to opioids, their risk for addiction, and how effectively they respond to **Naloxone** or **Naltrexone**.

- **Mu-Opioid Receptor Variations**: The mu-opioid receptor is primarily involved in the euphoric and pain-relieving effects of opioids. Variations in the genes that code for this receptor can make some individuals more susceptible to addiction or overdose. For instance, individuals with certain genetic variations may require higher doses of Naloxone for effective overdose reversal.

- **Naltrexone Response**: Naltrexone works by blocking the mu-opioid receptors to reduce cravings and prevent relapse in individuals with opioid and alcohol use disorders. However, individuals with specific **genetic profiles** may have a different response to Naltrexone, which could affect treatment outcomes. Personalized medicine could help identify these variations, ensuring that patients are given the right treatment at the right dose.

Biomarkers for Tailored Treatment Plans

Biomarkers are biological indicators that can be measured to assess the presence or progress of disease, or the effects of a treatment. As research advances, biomarkers may also be used to predict how patients will respond to **Naloxone** and **Naltrexone** treatments.

- **Pain and Addiction Biomarkers**: Researchers are currently working to identify biomarkers that could help determine how a patient will respond to opioid treatment or to medications like **Naloxone** and **Naltrexone**. These biomarkers could help physicians predict which patients are at higher risk of addiction or overdose and guide the selection of the most appropriate treatment strategies.

- **Response to Naltrexone**: Some individuals may experience side effects from Naltrexone, such as nausea or headache, while others may not respond at all. The identification of genetic or protein biomarkers associated with Naltrexone efficacy could help clinicians personalize treatment plans, improving the effectiveness of therapy and minimizing side effects.

Pharmacogenomics: The Future of Personalized Medicine in Endorphin Inhibition

Pharmacogenomics—the study of how genes affect a person's response to drugs—holds great promise for improving the use of **Naloxone** and **Naltrexone**. By examining how an individual's genetic makeup influences their reaction to these medications, researchers can develop **genetically tailored treatments** that are more effective and have fewer adverse effects.

For instance, genetic tests may soon be available that can determine how an individual metabolizes **Naltrexone** or how their opioid receptors respond to **Naloxone**, allowing for **precise dosing** and reducing the risk of treatment failure. This would mark a significant shift from the one-size-fits-all approach currently used in addiction treatment to a model where **treatment is individualized**, taking into account the patient's unique genetic and molecular profile.

Future Directions in Personalized Addiction Treatment

As personalized medicine continues to evolve, the role of **Naloxone** and **Naltrexone** in addiction treatment is likely to become more nuanced and specific to each individual. These drugs may become part of a larger, tailored treatment strategy that integrates genetic testing, **biomarkers**, and **personal health history** to maximize the chances of success.

- **Personalized Recovery Plans**: Rather than using a standard regimen of Naloxone and Naltrexone, clinicians may begin to combine these drugs with **other medications** or therapies that are tailored to the patient's unique characteristics. For example, certain patients may benefit from additional mood-stabilizing medications, while others may require cognitive-behavioral therapy (CBT) or **contingency management** strategies as part of their recovery plan.

- **Expanding Treatment Beyond Addiction**: As we have seen throughout this book, **Naloxone** and **Naltrexone** are valuable in treating not just opioid use disorder but also alcohol dependence, chronic pain, and other behavioral health conditions. The development of **personalized treatments** could lead to **novel interventions** for mental health disorders, allowing for a more precise and effective approach to treatment.

Conclusion: The Future of Endorphin Inhibition

The journey toward mastering **endorphin inhibition** and the clinical use of **Naloxone** and **Naltrexone** has only just begun. The future holds incredible potential for **personalized medicine** to transform the way we use these medications, improving outcomes for individuals struggling with addiction, chronic pain, and a variety of other health conditions.

By integrating advances in **genetic research**, **biomarkers**, and **pharmacogenomics**, the medical community can move toward a more personalized, effective approach to treatment that not only addresses the **underlying causes of addiction** and **pain** but also provides patients with a better quality of life and a greater chance at lasting recovery.

This personalized approach will ultimately redefine the role of **Naloxone** and **Naltrexone** in addiction treatment and beyond, making them even more powerful tools in the fight against the opioid crisis and other health challenges in the years to come.

Chapter 23: The Role of Naloxone and Naltrexone in Global Public Health

Introduction

The opioid crisis is a global public health emergency, with opioid use disorder (OUD) and opioid overdose deaths affecting millions of lives around the world. As countries struggle to combat this epidemic, Naloxone and Naltrexone have emerged as essential tools in reducing opioid-related harm and improving public health outcomes. From emergency overdose reversals to long-term recovery support, the role of these medications extends far beyond individual treatment—they are now key components of **global health strategies** aimed at mitigating the widespread effects of opioid addiction.

This chapter explores how **Naloxone** and **Naltrexone** are being incorporated into **public health initiatives** worldwide, as well as the challenges and successes faced in different regions. We will examine the importance of **global accessibility**, **policy frameworks**, and **community-based programs** that promote the use of these drugs...

Chapter 24: Integrating Naloxone and Naltrexone into Multidisciplinary Treatment Models

Introduction

The treatment of opioid use disorder (OUD), alcohol use disorder (AUD), and other conditions involving the opioid system is most effective when approached through a **multidisciplinary model**—one that integrates various healthcare professionals, treatment modalities, and support systems. While **Naloxone** and **Naltrexone** are critical components in the management of addiction and overdose prevention, their true therapeutic potential is fully realized when combined with comprehensive care that addresses the **physical**, **psychological**, and **social** aspects of recovery.

In this chapter, we explore how **Naloxone** and **Naltrexone** can be integrated into **multidisciplinary treatment models**, including **medical**, **psychosocial**, and **community-based** interventions. By examining these collaborative approaches, we will highlight the importance of...

Chapter 25: Conclusion: The Path Forward in Endorphin Inhibition and Public Health

Introduction

As we reach the conclusion of our journey through the science, applications, and future of **Naloxone** and **Naltrexone**, it becomes clear that the understanding and mastery of endorphin inhibition hold profound implications for the future of healthcare. These two medications, while primarily known for their role in treating opioid overdose and addiction, represent a broader opportunity to reshape how we approach addiction, pain management, and recovery. Their impact reaches beyond the individual, influencing **public health systems**, **policy**, and **global health efforts**.

This final chapter reflects on the key themes of this book and explores the way forward —how we can optimize the use of these medications to create more effective treatment models, address emerging public health concerns, and ultimately bring an end to the opioid epidemic. It also highlights the broader implications of endorphin regulation, from ethical considerations to the potential for new therapeutic approaches.

The Evolving Role of Naloxone and Naltrexone in Public Health

Over the past few decades, the role of **Naloxone** and **Naltrexone** in addiction and overdose prevention has transformed from niche solutions into cornerstone therapies in the battle against opioid use disorder (OUD). Yet, as we've seen throughout this book, the applications of these medications go beyond just **opioid-related issues**. **Naloxone**, as a **life-saving intervention**, has saved countless lives in emergency situations, and its presence in **overdose prevention programs** has become an essential element in the fight against opioid fatalities.

Naltrexone, with its capacity to help individuals sustain recovery and manage cravings in opioid and alcohol use disorders, is similarly expanding its reach. As both medications continue to evolve, they will increasingly become part of **multidisciplinary treatment models** that address the complexity of addiction in a holistic way.

However, the future of **Naloxone** and **Naltrexone** cannot be viewed solely through the lens of addiction treatment. Both drugs are pivotal in a larger **public health framework** that seeks to integrate **prevention**, **early intervention**, **treatment**, and **long-term recovery** into a **comprehensive strategy** to combat substance use disorders and improve the health and wellbeing of affected communities. By expanding access to these medications, advocating for **harm-reduction policies**, and **educating the public**, society can make significant strides in reducing opioid overdose deaths and facilitating long-term recovery.

The Road Ahead: Key Challenges and Opportunities

The challenges faced in the fight against the opioid epidemic are considerable. Access to Naloxone, the stigma associated with addiction treatment, and the persistent issues of overdose deaths all represent obstacles to achieving meaningful progress. However, these challenges also present opportunities for innovation, particularly in the areas of **policy reform**, **education**, and **healthcare delivery**.

1. **Access and Availability**: One of the major barriers to the effective use of **Naloxone** and **Naltrexone** is ensuring that they are widely accessible to those who need them most. Expanding access to Naloxone in **communities, hospitals, pharmacies**, and **first responders** is crucial. Furthermore, making **Naltrexone** available to those struggling with addiction, particularly in underserved areas, can be a game-changer.

2. **Education and Stigma Reduction**: A key challenge to widespread adoption is the stigma that still surrounds **addiction treatment**. Public education campaigns and training for healthcare providers and community organizations can go a long way toward reducing misconceptions and encouraging more people to seek the help they need. **Naloxone** is not just a **rescue drug**; it is a **preventive tool** that enables users to survive long enough to enter treatment and recovery.

3. **Innovative Treatment Models**: The future of addiction treatment lies in the integration of **Naloxone** and **Naltrexone** into more **comprehensive treatment regimens**. Research into their combined use with behavioral therapies, **mental health support**, and **lifestyle interventions** will allow for more personalized and effective approaches to long-term recovery.

4. **Global Public Health Impact**: Beyond national borders, **Naloxone** and **Naltrexone** have the potential to make a significant difference in **global health**. In countries experiencing rising rates of opioid addiction, **overdose deaths**, and **substance abuse-related health crises**, these medications can be part of a broader effort to reduce harm and provide effective treatment solutions. International partnerships and **global health initiatives** focused on opioid addiction and overdose prevention can help address the crisis on a global scale.

Looking to the Future: Endorphin Regulation and Beyond

While Naloxone and Naltrexone will remain central to our understanding of endorphin inhibition for the foreseeable future, the field of **opioid receptor research** is evolving. New discoveries about the interaction between **endorphins**, **opioid receptors**, and other **neurotransmitters** could lead to the development of more **targeted** and **effective** treatments for addiction, pain management, and even mental health conditions like depression and anxiety. Furthermore, the growing interest in **personalized medicine** and **genomic data** may provide insights into how **genetic factors** influence individual responses to these medications, paving the way for even more tailored treatments.

Conclusion: A Call to Action

Mastering endorphin inhibition through the use of **Naloxone** and **Naltrexone** is not just about managing opioid use disorder or preventing overdose deaths—it is about creating a **healthier, more resilient society**. By understanding the profound effects of endorphins on both the mind and body, we can begin to craft a more comprehensive and humane approach to addiction treatment and recovery.

The journey to mastering endorphin inhibition is ongoing, but with continued research, policy reform, and public health initiatives, we have the opportunity to end the opioid crisis, support those in recovery, and transform how we think about addiction treatment in the 21st century. The path forward lies in **collaboration**, **education**, and the unwavering commitment to providing the best possible care to those who need it most.